CAMBRIDGE LIBRARY COLLECTION

Books of enduring scholarly value

Technology

The focus of this series is engineering, broadly construed. It covers technological innovation from a range of periods and cultures, but centres on the technological achievements of the industrial era in the West, particularly in the nineteenth century, as understood by their contemporaries. Infrastructure is one major focus, covering the building of railways and canals, bridges and tunnels, land drainage, the laying of submarine cables, and the construction of docks and lighthouses. Other key topics include developments in industrial and manufacturing fields such as mining technology, the production of iron and steel, the use of steam power, and chemical processes such as photography and textile dyes.

The Atlantic Telegraph

In 1866, William Howard Russell (1820–1907) published this work, the official account of the July 1865 expedition on board the *Great Eastern* to lay a cable along the Atlantic Ocean floor between Valentia, Ireland, and Foilhummerum Bay in Newfoundland. It is illustrated with 26 lithographs of watercolours by Robert Dudley, who also travelled with the expedition. The cable, constructed by the Telegraph Construction & Maintenance Company, was designed to create a communications bridge between North America and Europe, enabling telegrams to be sent and received within minutes, when previously messages could be sent only by ship. The 1865 expedition was the fourth attempt to lay the cable, and although after 1200 miles the cable broke and was lost in the ocean, an expedition the following year was finally successful. This lively account of a pioneering attempt will appeal to anyone with an interest in the history of technology.

Cambridge University Press has long been a pioneer in the reissuing of out-of-print titles from its own backlist, producing digital reprints of books that are still sought after by scholars and students but could not be reprinted economically using traditional technology. The Cambridge Library Collection extends this activity to a wider range of books which are still of importance to researchers and professionals, either for the source material they contain, or as landmarks in the history of their academic discipline.

Drawing from the world-renowned collections in the Cambridge University Library, and guided by the advice of experts in each subject area, Cambridge University Press is using state-of-the-art scanning machines in its own Printing House to capture the content of each book selected for inclusion. The files are processed to give a consistently clear, crisp image, and the books finished to the high quality standard for which the Press is recognised around the world. The latest print-on-demand technology ensures that the books will remain available indefinitely, and that orders for single or multiple copies can quickly be supplied.

The Cambridge Library Collection will bring back to life books of enduring scholarly value (including out-of-copyright works originally issued by other publishers) across a wide range of disciplines in the humanities and social sciences and in science and technology.

The Atlantic Telegraph

William Howard Russell

CAMBRIDGE
UNIVERSITY PRESS

CAMBRIDGE UNIVERSITY PRESS

Cambridge, New York, Melbourne, Madrid, Cape Town,
Singapore, São Paolo, Delhi, Tokyo, Mexico City

Published in the United States of America by Cambridge University Press, New York

www.cambridge.org
Information on this title: www.cambridge.org/9781108072472

This edition first published 1866
This digitally printed version 2011

ISBN 978-1-108-07247-2 Paperback

The Atlantic Telegraph

William Howard Russell

CAMBRIDGE
UNIVERSITY PRESS

CAMBRIDGE UNIVERSITY PRESS

Cambridge, New York, Melbourne, Madrid, Cape Town,
Singapore, São Paolo, Delhi, Tokyo, Mexico City

Published in the United States of America by Cambridge University Press, New York

www.cambridge.org
Information on this title: www.cambridge.org/9781108072472

This edition first published 1866
This digitally printed version 2011

ISBN 978-1-108-07247-2 Paperback

The Atlantic Telegraph

William Howard Russell

CAMBRIDGE UNIVERSITY PRESS

Cambridge, New York, Melbourne, Madrid, Cape Town,
Singapore, São Paolo, Delhi, Tokyo, Mexico City

Published in the United States of America by Cambridge University Press, New York

www.cambridge.org
Information on this title: www.cambridge.org/9781108072472

© in this compilation Cambridge University Press 2011

This edition first published 1866
This digitally printed version 2011

ISBN 978-1-108-07247-2 Paperback

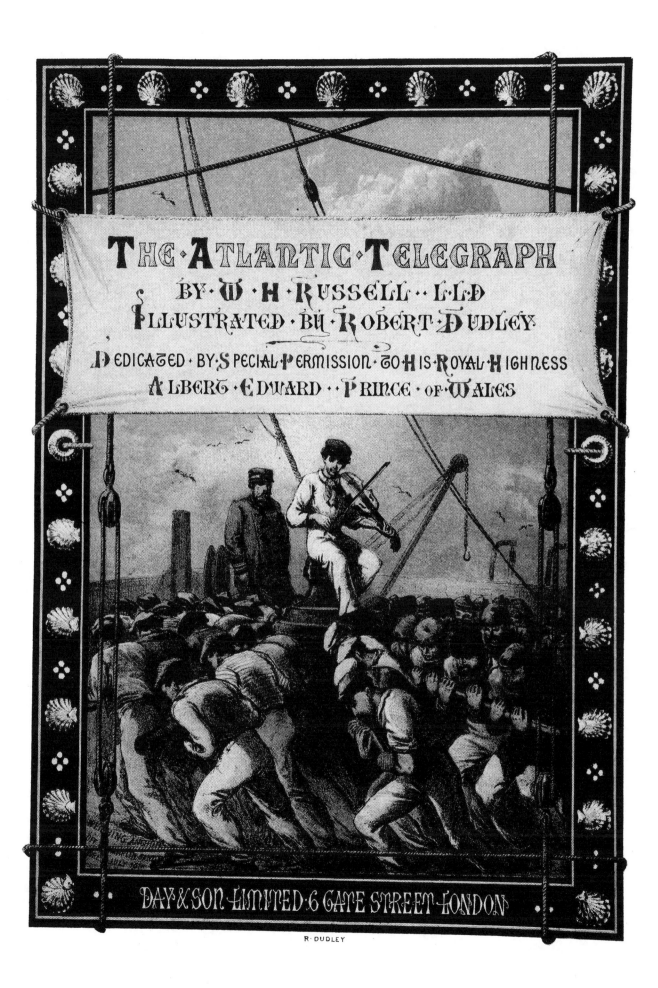

LIST OF ILLUSTRATIONS.

—•—

THE

ATLANTIC TELEGRAPH.

I SHALL not detain the readers of this brief narrative with any sketch of the progress of electrical science. There are text-books, cyclopædias, and treatises full of information concerning the men who worked in early days, and recording the labours of those who still toil on, investigating the laws and developing the applications of the subtle agency which has long attracted the attention of the. most acute, ingenious, and successful students of natural philosophy. For the last two centuries the greater number of those whose names are known in science have made electrical experiments a favourite pursuit, or turned to them as an agreeable recreation from severer studies. The rapidity with which electricity travels for considerable distances through insulated conductors soon suggested its use as a means of transmitting intelligence; but the high tension of the currents from friction machines, and the difficulty of insulating the conductors, were practical obstacles to the employment of the devices, some of them ingenious, recommended for that purpose from year to year. Otto Von Guericke, and his globe of sulphur; Grey, with his glass tube and ·silken cords; and Franklin, with his kite, were, however, the precursors of the philosophers who have done much, and whose successors may yet do much more, for the world. It is not easy to decide whether it is the man who gives a new idea to the world, or he who embodies that idea in a form and turns it into a fact, who is deserving of the credit to be assigned to any invention. A vague expression of belief that a certain end may be attained at a future period by means then unknown does not constitute a discovery, and does not entitle the person who utters it, verbally or in writing, to the honour which is due to him who indicates

B

specifically the way of achieving the object, or who actually accomplishes it by methods he has either invented or applied. The Marquis of Worcester certainly did not invent the steam-engine ; neither did Watson, Salva, Sœmmering, or Ronalds, or any other of the many early experimentalists, discover electric telegraphy. But there is a degree of credit due to those who, contending with imperfect materials and want of knowledge, persist in working out their ideas, and succeed in rescuing them from the region of chimæras. The inventions of one render capable of realisation the ideas of another, which but for them had remained dreams and visions. The introduction of a novel product into commerce, or the chance discovery of some property in a common material, may draw a project out of the limbo of impracticabilities. A suggestion at one period may be more valuable than an invention at another, and adaptations may be more useful than discoveries. Indeed, when the testimony on which men's reputations, as finders or makers, rest, is critically examined, a suspicion is often generated that there have been many Vespuccis in the world who have given names to places they never found, and taken or received credit for what they never did.

If any person takes an interest in determining who was the inventor of electric telegraphy, he should study the works and mark the improvements of the natural philosophers of the last as well as of the present, century, and he can then arrive at some result without exciting national jealousy, or injuring individual suscepti- bilities. Humboldt assigns the credit of making the first electric telegraph to Salva, who constructed a line 26 miles long, from Madrid to Aranjuez, in 1798. Russia claims the honour of having invented aerial telegraphic lines, because Baron Von Schilling proposed a line for the Emperor from St. Petersburg to Peterhoff, below Cronstadt, in 1834, and was laughed at by scientific Muscovites for his pains. But the Baron certainly did transmit messages along wires supported by poles in the air. The Count du Moncel, in his recent " Traité de Télégraphie Electrique," gives to Mr. Wheatstone the palm as the original inventor of submarine Cables, to which award, no doubt, there will be some dissent. Mr. Wheatstone, however, as early as 1840, brought before the House of Commons the project of a cable, to be laid between Dover and Calais, though he does not seem to have had at the time any decided views as to the mode in which insulation was to be obtained. In 1843, Professor Morse, detailing the results of some experiments with an electric magnetic telegraph between Washington and Baltimore, in a letter to the Secretary of the United States, wrote : " The practical inference from this law is that a telegraphic communication on the electric-magnetic plan, may with certainty be established across the Atlantic Ocean. Startling as this may seem now, I am confident the time will come when this project will be realised." But for the experiments and discoveries of Oersted, Sturgeon, Ampére, Davy, Henry, and Faraday, and a long list of others, such suggestions would have remained as little likely to be realised as the Bishop of Llandaff's notions of a flying-machine, or the crude theories of the alchemists. He who first produces a practical result—something which, however imperfect, gives a result to be seen and felt, and appreciated by the senses—is the true ποιήτης—the maker and inventor, whom the world should recognise, no matter how much may be done by others to improve his work, each of those improvers being, after his kind, deserving of recognition for what he does. A year before Professor Morse wrote the letter to Mr. Spencer, he took some steps to show that which he prophesied was practicable. In the autumn of the year 1842 he stretched a submarine cable from Castle Garden to Governor's Island in the harbour of New York, demonstrated to the American Institute the possibility of effecting electric communication through the sea, and submitted that telegraphic communication might with certainty be established across the Atlantic. Later in the same year he sent a current across the canal at Washington. But that was not the first current transmitted under water, for as early as 1839, Sir W. O'Shaughnessy, the late Superintendent of Electric Telegraphs in India, hauled an insulated wire across the Hooghly at Calcutta, and produced electrical phenomena at the other side of the river. In 1846, Col. Colt, the patentee of the revolver, and Mr. Robinson, of New York, laid a wire across the river from New York to Brooklyn, and from Long Island to Coney Island. In 1849, Mr. Walker sent messages to shore through two miles of insulated wire from a battery on board a steamer off Folkestone.

It was in 1851 that an electric cable was actually laid in the open sea, and worked successfully; and the wire which then connected Dover with Calais was beyond question the first important line of submarine telegraph ever attempted. In the year 1850, Mr. Brett obtained a concession from the French Government for effecting this object,—an object regarded at the time as one purely chimerical, and decried by the press as a gigantic swindle. The cable which was made for the purpose consisted of a solid copper wire, covered with gutta percha. When tested by Mr. Woollaston, it was found to be so imperfect from air holes in the gutta-percha, that the water found its way to the copper wire,—an imperfection which was however shortly repaired. This cable was manufactured at the Gutta Percha works, on the Wharf Road, City Road, under the superintendence of the late Mr. Samuel Statham; was then coiled on a drum, and conveyed by steam-tug to Dover, and in the year 1850 was payed out from Dover to Calais. The landing-place in France was Cape Grisnez, from which place a few messages passed, so as to comply with the terms of the concession and test the accuracy of the principle. The communication thus established between the Continent and England was, after a few hours, abruptly stopped. A diligent fisherman, plying his vocation, took up part of the cable in his trawl, and cut off a piece, which he bore in triumph to Boulogne, where he exhibited it as a specimen of a rare seaweed, with its centre filled with gold. It is believed that this "pescatore ignobile" returned again and again to search for further specimens of this treasure of the deep : it is, at all events, perfectly certain that he succeeded in destroying the submarine cable.

This accident caused the attention of scientific men to be directed to the discovery of some mode of preserving submarine cables from similar casualties, and a suggestion was made by Mr. Küper, who was engaged in the manufacture of wire ropes, to Mr. Woollaston and to Mr. T. R. Crampton, that the wire insulated with gutta-percha should form a core or centre to a wire rope, so as to give protection to it during the process of paying out and laying down, as well as to guard it from the anchors of vessels and the rocks, and to secure a perfect electrical continuity.

Mr. Crampton, who had already accepted the contract for laying the cable between England and France, and had given up much of his time to the study of the subject, adopted this idea, and in 1851 he and several gentlemen associated for the purpose laid the cable between Dover and Calais, where it has since remained in perfect order, constituting the great channel of electrical communication between England and the Continent. It was made by Wilkins & Weatherly, Newall & Co., Küper & Co., and Mr. Crampton. The exertions of the last-named eminent engineer in laying the first cable under water, and his devotion to an object towards which he largely contributed in money, are only known to a few, and have never been adequately acknowledged.

The success of that form of cable having been thus completely established, several lines of a similar character were laid during the following years between England and Ireland and parts of the Continent : one, 18 miles long, across the Great Belt, made by Newall & Co. ; one from Dover to Ostend, by the same makers and by Küper & Co. ; one from Donaghadee to Portpatrick, by Newall & Co. ; one from Holyhead to Howth ; and one from Orfordness to the Hague.

The superiority of a line with wire-rope cover to other descriptions of cable was illustrated in 1853. At that period the Electric and International Telegraph Company determined upon laying down four wires between England and the Continent, but they rejected the heavy cable, and adopted the suggestion of their engineer to use four separate cables of light wire. The cost of maintaining these light cables from injury by anchors, &c., was so great that they were picked up, and heavy cables of great strength were substituted, which have given no trouble or anxiety, and have always been in good order.

The Old World had twelve lines of submarine cable laid ere the United States turned their attention to the uses of such forms of telegraph. Italy had been connected with Corsica by a line 110 miles long, and Denmark had joined one of her little islands to the other, ere the Great Republic gave a thought to the matter. But there were excuses for such indifference. The Telegraphic system, to which Morse, Bain, House, and others, had given such development, although the first line was not constructed till 1844, extended rapidly all over the vast extent of the Atlantic and Gulf States. The people were on the same continent, the land was all their own, their greatest rivers could be traversed by wires; and so it was that, whilst Mr. Morse was engaged in protecting his patents, and the Americans, self-contained, were not looking beyond the limits of their shores, a British North American Province took the first step which was made at the other side of the Atlantic to lay down a submarine cable. In 1851-2 a project was started in Newfoundland, to run a line of steamers between Galway and St. John's in connection with a telegraph to Cape Ray, where a submarine Cable was to be laid to Cape Breton, and thence the news was to be carried by means of another cable from New Brunswick to Prince Edward's Island. The Roman Catholic Bishop of Newfoundland is stated to have been the original proposer of a scheme for connecting the island with the United States, but the credit of actually laying down the first submarine cable at the other side of the Atlantic belongs to Mr. F. N. Gisborne, an English engineer. He had been previously engaged in the telegraph department at Montreal, and had some knowledge of the subject, but he happened to be in London at the time of Brett's success. On his return to America he applied himself to get up a Company for the purpose of facilitating telegraphic communication between Europe and the United States. After much difficulty the Company was formed, and an Act was passed by the Legislature of Newfoundland, in 1852, conferring the important privileges upon it, in event of the completion of the project in Newfoundland, which are now possessed by the Atlantic Telegraph Company. Mr. Gisborne was superintendent and engineer of the Company, and he set to work with energy to construct a road from St. John's to Cape Ray, over a barren and resourceless tract of 400 miles, and made a survey of the coast line, during which he was exposed to great hardships. He succeeded at last in laying an insulated cable, made by Newall & Co., from New Brunswick to Prince Edward's Island across the Straits of Northumberland, 11 miles long, in

22 fathoms of water; but was not successful in a similar attempt to connect Newfoundland with Cape Breton. Meantime the Company became involved in pecuniary difficulties, and Mr. Gisborne, early in 1854, on the suspension of the works, proceeded to New York, where he hoped to find money to enable him to carry out the telegraphic scheme among the keen speculators and large-pursed merchants. Through an accidental conversation at the hotel in which he was staying, he obtained an interview with Mr. Cyrus Field. He laid his plans before that gentleman, who had no desire to resume an active career, having just returned from travelling in South America, with the intention of enjoying the fortune his industry and sagacity had secured ere he had arrived at the middle term of life. But Mr. Field listened to Mr. Gisborne with attention, and then began to think over the project—"To lay these submarine cables so as to connect Newfoundland with Maine?—Good. To run a line of steamers from St. John's to Galway?—Certainly. It would shorten the time of receiving news in New York from Europe four or five days." And so the brain worked and thought. Then suddenly, "But if a cable can be laid in the bed of these seas—if the Great Atlantic itself could be spanned?" Here was an idea indeed. Deep and broad seas had been traversed in Europe, but here was one of the great oceans of the world, of depth but faintly guessed at, and of nigh 2000 miles span from shore to shore! Would it be within the limits of human resources to let down a line into the watery void, and to connect the Old World with the New? What a glorious thought! Was it a vision, or was it one of those inspirations from which originate grand enterprises and results which change the destinies of the world? Mr. Field terminated his reflections that night by an eminently practical measure. Ere he retired to rest he sat down and wrote two letters,—one to Lieut. Maury, U.S.N., to ask his opinion concerning the possibility of laying down a cable in the bottom of the Atlantic; the other to Professor Morse, to inquire whether he thought it practicable to send an electric current through a wire between Europe and America. Lieut. Maury, in answering in the affirmative, wrote, "Curiously enough, when your letter came I was looking over my letter to the Secretary of the Navy on that very subject." And, in fact, on the 22nd February, 1854, Lieut. Maury made a long communication to Mr. Dobbin, Secretary, United States Navy, from the Observatory, Washington, respecting a series of deep-sea soundings made by Lieut. Berryman, U.S.N., brig Dolphin, from Newfoundland to Ireland, in connection with researches on the winds and currents, carried on for the National Observatory. It is obvious that Lieut. Maury, as well as many others probably, had thought of the same idea as Mr. Field. He says, "The result is highly interesting, in so far as the bottom of the sea is concerned, upon the

question of a submarine telegraph across the Atlantic ;" and he goes on to make it the subject of a special report, in which occurs the following passages ;—

"This line of deep-sea soundings seems to be decisive of the question as to the practicability of a Submarine Telegraph between the two continents, in so far as the bottom of the deep sea is concerned. From Newfoundland to Ireland, the distance between the nearest points is about 1,600 miles ;* and the bottom of the sea between the two places is a plateau, which seems to have been placed there especially for the purpose of holding the wires of a Submarine Telegraph, and of keeping them out of harm's way. It is neither too deep nor too shallow ; yet it is so deep that the wires, but once landed, will remain for ever beyond the reach of vessels' anchors, icebergs, and drifts of any kind, and so shallow that the wires may be readily lodged upon the bottom. The depth of this plateau is quite regular, gradually increasing from the shores of Newfoundland to the depth of from 1,500 to 2000 fathoms as you approach the other side. The distance between Ireland and Cape St. Charles, or Cape St. Lewis, in Labrador, is somewhat less than the distance from any point of Ireland to the nearest point of Newfoundland. But whether it would be better to lead the wires from New-foundland or Labrador is not now the question ; nor do I pretend to consider the question as to the possibility of finding a time calm enough, the sea smooth enough, a wire long enough, a ship big enough, to lay a coil of wire 1,600 miles in length ; though I have no fear but that the enterprise and ingenuity of the age, whenever called on with these problems, will be ready with a satisfactory and practical solution of them.

" I simply address myself at this time to the question in so far as the bottom of the sea is concerned, and as far as that the greatest practical difficulties will, I apprehend, be found after reaching soundings at either end of the line, and not in the deep sea. * * Therefore, so far as the bottom of the deep sea between Newfoundland, or the North Cape, at the mouth of the St. Lawrence, and Ireland, is concerned, the practicability of a Submarine Telegraph across the Atlantic is proved."

Professor Morse, in 1843, indicated his conviction that a magnetic current could be conveyed across the Atlantic, and his reply to Mr. Field was now given with increased confidence to the same effect. Thus encouraged, Mr. Field took measures to form a Company to purchase the rights of the Newfoundland Company, and to connect Newfoundland with Ireland by means of a submarine telegraph across the Atlantic. He entered into an agreement with Mr.

* "From Cape Freels, Newfoundland, to Erris Head, Ireland, the distance is 1,611 miles ; from Cape Charles, or Cape St. Lewis, Labrador, to ditto, the distance is 1,601 miles."

Gisborne for the purchase of the privileges of the Company for 8000*l.*, under certain conditions.   Then he put down the names of ten of the principal capitalists in New York, and proceeded to unfold his project to each in succession ; and having secured the adhesion of Mr. Cooper, Mr. Taylor, Mr. Roberts, Mr. White, and the advice of his brother, Mr. D. Field, he called a meeting of these gentlemen at his house on 7th March.   Similar meetings took place at his residence on 8th, 9th, and 10th, and after full discussion and consideration it was resolved to form "The New York, Newfoundland, and London Telegraph Company," of which Peter Cooper was President ; Moses Taylor, Treasurer ; Cyrus Field, C. White, M. O. Roberts, Directors ; and D. D. Field, Counsel.   Mr. C. Field, his brother, and Mr. White were commissioned to proceed to Newfoundland, to obtain from the Legislature an act of incorporation, and set out for that purpose on March 15th. On their arrival at St. John's, the Governor convoked the Executive Council.   He also sent a special message to the Legislature, then in session, recommending them to pass an act of incorporation, with a guarantee of interest on the Company's bonds to the amount of 50,000*l.*, and to make them a grant of fifty square miles of land on the island of Newfoundland, conditional on the completion of the Telegraph.

After some little delay, the Legislature, with one adverse member only, granted the valuable privileges to the Company which were subsequently transferred to the Atlantic Telegraph Company.   They constitute, in fact, a monopoly of telegraphic rights in Newfoundland, the value of which was enhanced afterwards by similar concessions from the state of Maine, Nova Scotia, Prince Edward's Island ; and liberal encouragement from Canada.   There is much to be said against concessions, and monopolies, and patents, on abstract grounds ; but it is quite clear that in certain circumstances men will not venture money and spend time, without the prospect of the ulterior advantages such protection is calculated to ensure.   The Government has, however, informed Colonial and Provincial Legislatures that in future Her Majesty will be advised not to give her ratification to the creation of similar monopolies.   By their chartered rights the new Company obtained the exclusive privilege for fifty years of landing cables on Newfoundland and Labrador, which embraces a coast extending southwardly to Prince Edward's Island, Cape Breton, Nova Scotia, the State of Maine, and their respective dependencies ; and westwardly to the very entrance of Hudson's Straits.   The Company also secured a grant of fifty square miles of land on the completion of Telegraph to Cape Breton ; a similar concession of additional fifty square miles when the Cable shall have been laid between Ireland and Newfoundland ; a guarantee of interest for twenty years at 5 per cent. on 50,000*l.* ; a grant of 5000*l.* in money towards building a road along the line of the Telegraph ; and the remission of duties on the importation of all wires and materials for the use of the Company.

The Company also obtained from the Legislature of Prince Edward's Island, in May, 1854, the exclusive privilege for fifty years of landing cables on the coast ; a free grant of one thousand acres of land ; and a grant of 300*l.* currency per annum for ten years.

From Canada the Company obtained an Act authorising the building of telegraph lines throughout the Provinces, accompanied by the remission of duties on all wires and materials imported for the use of the Company.

Nova Scotia, in 1859, gave the Company a grant of exclusive privilege, for twenty-five years, of landing telegraphic cables from Europe on the shores of the Province.

The State of Maine accorded the Company a grant of the exclusive privilege, for twenty-five years, of landing European telegraph cables on the seaboard.

From Great Britain eventually the Company obtained an annual subsidy of 14,000*l.* sterling until the net profits of the Company should reach 6 per cent. per annum, on the whole capital of 350,000*l.* sterling, the grant to be then reduced to 10,000*l.* sterling per annum, for a period of twenty-five years ; two of the largest steamships in the navy to lay the cable, and two steamers to aid them ; and a careful examination of the soundings by vessels of the Royal Navy.

From the United States the Company obtained an annual subsidy of $70,000 until the net profits yielded 6 per cent. per annum, then to be reduced to $50,000 per annum, for a period of twenty-five years, subject to termination of contract by Congress after ten years, on giving one year's notice. The United States government also granted the steamship Arctic to make soundings, and steam-ships Niagara and Susquehanna to assist in laying the cable. A government steamer was also ordered to make further soundings on the coast of Newfoundland.

Long ere the Company had been placed in possession of such beneficial rights, and obtained such a large amount of favour, Mr. Field, who threw every energy of body and mind into the work, and was entrusted by his brother directors with the general management of affairs, proceeded to carry out the engagements the Company had entered into with the local legislatures. It has been said that the greatest boons conferred on mankind have been due to men of one idea. If the laying of the Atlantic Cable be among these benefits, its consummation may certainly be attributed to the man who, having many ideas, devoted himself to work out one idea with a gentle force and a patient vigour which converted opposition and overcame indifference. Mr. Field may be likened either to the core, or to the external protection, of the Cable itself. At times he has been its active life; again he has been its iron-bound guardian. Let who will claim the merit of first having said the Atlantic Cable was possible, to Mr. Field is due the inalienable credit of having made it possible, and of giving to an abortive conception all the attributes of healthy existence.

The first step in the great enterprise, now fairly inaugurated, was the connection of St. John's with the telegraphic lines already in operation in Canada and the United States.

Mr. Field was despatched to England, as there were no firms established for the manufacture of submarine cables in the United States, to order the necessary work to be done, and to raise money. He previously ordered specimens of cable to be made, so that when he landed in England they were ready for his inspection; and soon after his arrival he entered into a contract with Messrs. Küper & Co. (subsequently Glass, Elliot, & Co.) for a cable to be laid across the Gulf of St. Lawrence. He held interviews with eminent engineers and electricians, among whom were Mr. Brunel, Mr. (now Sir C.) Bright, Mr. Brett, and Mr. Whitehouse, respecting his larger project, which led to extended and valuable experiments. The cable for Newfoundland was formed in three strands, and had three conducting wires; and Mr. Field undertook to lay it, under the direction of Mr. Canning. In August, 1855, the first attempt was made; but off Cape Ray a violent gale arose, and it was deemed necessary by the master of the vessel to cut the cable. This disappointment was not in the least a discouragement. Another contract was made by Mr. Field with Messrs. Küper & Co. to make and lay a cable at their own risk, which was executed by Mr. Canning in the Propontis the following year. The station is at Point-au-Basque, near the western extremity of Newfoundland, and the telegraph runs across the island to Trinity Bay.

The opportunities for scientific experiments afforded by the manufacture of these cables were not neglected. The possibility of transmitting signals under water without fatal loss of power from the increased length of circuit was the first fact determined. The attention of the experimentalists was then directed to ascertain whether, having regard to existing theories, it would be possible to carry even a single conductor across the Atlantic without the aid of a cable so ponderous and so costly as to render it useless in a commercial point of view. A series of direct experiments were at once undertaken, which resulted in the establishment of the following facts :—first, that retardation of movement, in consequence of increasing distance, did not occur at a rate which could seriously affect a cable across the Atlantic; secondly, that increased dimensions in insulated marine conductors augmented the difficulties in obtaining velocity, so that bulk in a cable would not be requisite ; and, thirdly, that a velocity and facility which would satisfy all mere commercial and financial requirements in a line crossing the Atlantic, might be attained in the largest circuits. The next step was to actually make signals through 2000 miles of wire. This was accomplished through the kindness of the directors of the English and Irish Magnetic Company, who placed at the disposal of the experimentalists 5000 miles of under-ground wire. On the 9th of October, 1856, in the quiet of the night time, the experiment was tried successfully. Signals were distinctly and satisfactorily telegraphed through 2000 miles of wire, at the rate of 210, 241, and 270 per minute.

There was still a matter of the last importance to be determined. Was the state of the bed of the Atlantic really such as to warrant the conclusion that a wire 2000 miles long could be deposited and remain there without injury ?

Mr. Field, in order to ascertain this fact, obtained from the government of America the assistance of Lieut. Berryman, U.S.N., in the steam-ship Arctic, who succeeded, in July, 1856, in taking soundings across the Atlantic at distances varying from 30 to 50 miles, and, by means of scoops, or quills, bringing up specimens of the bottom, which, upon microscopic examination, proved to be composed of fine shells and sand.

As capital was needed for the execution of the enterprise which the confidence of moneyed men in the United States did not induce them to supply, and as it was desirable to enlist the support of the capitalists of Great Britain, Mr. Field was now authorised to form a company, with branches in both countries. Having secured the services of Mr. Brett, Mr. (now Sir C.) Bright, Mr. Woodhouse, and others, on the 1st of November, 1856, as Vice-President of the New York, Newfoundland, and London Telegraph Company, he issued an elaborate, able, and argumentative circular in London, headed, "Atlantic Telegraph," and made a tour through the great towns, addressing meetings in support of the project.

On the 6th of November, 1856, the prospectus was issued, with a nominal capital of 350,000l., represented by 350 shares of 1000l. each, and within one month the entire of the capital had been subscribed for, and the first instalment of 70,000l. paid up.

One hundred and six shares were taken in London, eight-eight in the United States, eighty-six in Liverpool, thirty-seven in Glasgow, and the remainder in other parts of England. Mr. Field stood as subscriber of 88,000l., and represented all America.

But it was not only from the public of Great Britain the project met encouragement. Ere the new company was formed, Mr. Field (13th September,

1855) addressed Lord Clarendon, requesting aid, and protection and privileges, and on the 20th November received a reply from the Secretary to the Treasury, engaging to furnish ships for soundings, and to consider favourably any request for help in laying the Cable, to pay 14,000*l*. (4 per cent. on capital) as remuneration for Government messages, till the net profits were 6 per cent., when the payment was to become 10,000*l*. for twenty-five years, and the Royal assent was given to the Act of Incorporation of the Company July 27th, 1857.

Mr. Field received far more encouragement in Great Britain, in Parliament and out of it, than he did at home. His bill was nearly rejected in the United States Senate, and it is stated only twenty-seven shares of the first stock were at first subscribed for in the States. On the motion of Mr. Seward, a resolution was passed in the Senate, United States, on the 23rd December, in compliance with which the President transmitted a copy of an application from the New York Office of the New York, Newfoundland, and London Telegraph Company, dated December 15th, in which the Directors set forth " their earnest desire to secure for the United States Government equal privileges with those stipulated for by the British Government in a work prosecuted thus far with American capital," and then recounted the terms agreed to by the Lords of the Treasury. On January 9th, 1857, Mr. Seward introduced a bill in the Senate to give and receive precisely the same privileges on the part of the United States Government. It was violently opposed, was only carried by one vote, and was not approved till March 3rd following.

The money being now forthcoming, the Provisional Directors of the Company proceeded to order the Atlantic Cable. Mr. Field was anxious that the order should be given to the firm which had manufactured the St. Lawrence Cable, but the Board thought it would be better to divide the contract, and on the 6th December, 1856, they entered upon agreements with the Gutta Percha Company for the supply of 2,500 miles of core, consisting of copper wire, with a triple covering of insulating substance, at 40*l*. per mile ; and also with Messrs. Glass, Elliot, & Co., of East Greenwich, and Messrs. Newall & Co., of Birkenhead, respectively, for the supply from each of 1,250 miles of the completed Cable for 62,000*l*. Within six months from that day, namely, on the 6th of July, 1857, the entire Cable was completed.

The policy of dividing the contract for the manufacture of the Cable was questioned at the time. When one portion of the Cable was to be made at East Greenwich and the other at Birkenhead, how was it possible that there could be any uniformity of supervision, any integrity of design, or any individual responsibility ? Again, how was it possible that the textile strength or conducting power of the Cable could be tested as satisfactorily as would have been the case were its manufacture entrusted to one firm? And, as it happened, the twist ran from right to left in one half, and from left to right in the other half of the Cable.

Before the prospectus was issued, every attention was paid that the characteristics of the Cable should be suited to its work; that it should not be too dense, lest its weight should render it unmanageable in the sea—nor too light, lest it should be at the mercy of the currents as it went down. It was decided that it should weigh a ton per mile, should be just so much heavier than the water which it displaced in sinking, and of such structure as could be easily coiled and yet be a rigid line, while its centre should be composed of wire capable of conveying electrical symbols through an extent of more than 2000 miles, and should retain complete insulation when immersed in the ocean. It was a subject of close and anxious inquiry how to obtain a Cable of this form and character. No fewer than sixty-two different kinds of rope were tested before one was determined on.

In the Cable finally adopted, the central conducting wire was a strand made up of seven wires of the purest copper, of the gauge known in the trade as No. 22. The strand itself was about the sixteenth of an inch in diameter, and was formed of one straightly drawn wire, with six others twisted round it; this was accomplished by the central wire being dragged from a drum through a hole in a horizontal table, while the table itself revolved rapidly, under the impulse of steam, carrying near its circumference six reels or drums each armed with copper wire. Every drum revolved upon its own horizontal axis, and so delivered its wire as it turned. This twisted form of conducting wire was first adopted for the rope laid across the Gulf of St. Lawrence in 1856, and was employed with a view to the reduction to the lowest possible amount of the chance of continuity being destroyed in the circuit. It seemed improbable in the highest degree that a fracture could be accidentally produced at precisely the same spot in more than one of the wires of this twisted strand. All the seven wires might be broken at different parts of the strand, even some hundreds of times, and yet its capacity for the transmission of the electric current not destroyed, or reduced in any inconvenient degree. The copper used in the formation of these wires was assayed from time to time during the manufacture to insure absolute homogeneity and purity. The strand itself, when subjected to strain, stretched 20 per cent. of its length without giving way, and indeed without having its conducting power much modified or impaired.

The copper strand of the Cable was rolled up on drums as it was completed, and was then taken from the drums to receive a coating of three separate layers of refined gutta percha ; these brought its diameter up to about three-eighths of an inch. The coating of gutta percha was made unusually thick, for the sake of diminishing the influence of induction, and in order that the insulation might be rendered as perfect as possible. This latter object was also furthered by the several layers of the insulating material being laid on in succession ; so that if there were accidentally any flaw in the one coat, the imperfection was sure to be removed when the next deposit was added. To prove the efficacy of the proceeding, a great number of holes were made near together in the first coating of a fragment of the wire, and the second coat was then applied in the usual way. The insulation of the strand was found to be perfect under these circumstances, and continued so even when the core was subjected to hydraulic pressure, amounting to five tons on the square inch. The gutta percha which was employed for the coating of the conducting strand, was prepared with the utmost possible care. Lumps of the crude substance were first rasped down by a revolving toothed cylinder, placed within a hollow case, the whole piece of apparatus somewhat resembling the agricultural turnip machine in its mode of action. The raspings were then passed between rollers, macerated in hot water, and well churned. They were next washed in cold water, and driven at a boiling-water temperature, by hydraulic power, through wire-gauze sieves, attached to the bottom of wide vertical pipes. The gutta percha came out from the sieves in plastic masses of exceeding purity and fineness, and those masses were then squeezed and kneaded for hours by screws, revolving in hollow cylinders, called masticators ; this was done to get the water out, and to render the substance of the gutta percha sound and homogeneous everywhere. At each turn of the screw, the plastic mass protruded itself through an opening left for feeding in the upper part of the masticator, and was then drawn back as the screw rolled on. When the mechanical texture of the refined mass was perfected by masticating and kneading, it was placed in horizontal cylinders, heated by steam, and squeezed through them by screw pistons, driven down by the machinery very slowly, and with resistless force. The gutta percha emerged, under this pressure, through a die, which received the termination of both cylinders, and which at the same time had the strand of copper wire moving along through its centre. The strands were drawn by revolving drums between the cylinders, and through the die. They entered the die naked bright copper wire, and issued from it thick, dull-looking cords, a complete coating of gutta percha having been attached to them as they traversed the die. Six strands were coated together, ranging along side by side at the first covering. Then a series of three lengths of the strand received the second coat together. The third coat was

F. Jones lith from a drawing by R. Dudley.

London Day & Son Limited Lith.

THE REELS OF GUTTA PERCHA COVERED CONDUCTING WIRE CONVEYED INTO TANKS AT THE WORKS AT GREENWICH

communicated to a solitary strand. The strand and its triple coating of gutta percha were together designated "the core."

The copper strand was formed and coated with gutta percha in two mile lengths. Each of these lengths, when completed, was immersed in water, and then carefully tested to prove that its continuity and insulation were both perfect. The continuity was ascertained by passing a voltaic current of low power through the strand from a battery of a single pair of plates, and causing it to record a signal after issuing from the wire. A different and very remarkable plan was adopted to determine the amount of insulation. One pole of a voltaic battery, consisting of 500 pairs of plates, was connected with the earth; the other pole was united to a wire which coiled round the needle of a very sensitive horizontal galvanometer, and then ran on into the insulated strand of the core, the end of which was turned up into the air, and left without any conducting communication. If the insulation was perfect, the earth would form one pole of the battery, and the end of the insulated strand the other pole, and the circuit be quite open and uninterrupted; consequently no current would pass, and the needle of the galvanometer would not be deflected in the slightest degree. If on the other hand there was any imperfection, or permeability in the sheath of gutta percha, a portion of the electricity would force its way from the strand through the faulty places and surrounding water to the earth, a current would be set up, and the needle of the galvanometer deflected; the deflection being in proportion to the current which passed, and therefore its degree would become a measure of the amount of imperfection.

When about fifty of the two-mile lengths of core were ready, these were placed in the water of the canal which ran past the gutta percha works, and were joined up by their ends into one continuous strand of 100 miles, the joints being covered with gutta percha. The hundred-mile length was then put through a careful scrutiny in the same way that the smaller portions were tried,—and next it was halved, quartered, and separated into groups of twenty, ten, and finally two miles, and each of these were again separately examined, and tested in comparison with similar lengths previously approved.

Whenever separate lengths of the gutta percha covered core were to be joined together, the gutta percha was scraped away for a short distance from the ends, and these were made to overlap. A piece of copper wire was then attached by firm brazing, an inch or two beyond the joint on one side, tightly bound round until it reached to the same extent on the other side, and then was there firmly brazed on again. A second binding was next rolled over the first in the same fashion, and extended a little way beyond it, and finally several layers of gutta percha were carefully laid over, and all round the joint by the agency of hot irons. If the core on each side of the joint was dragged opposite ways until the joint yielded, the outer investment of the wire unrolled spirally as the ends were pulled asunder, and so the conducting continuity of the strand was maintained, although the mechanical continuity of the strand itself was broken.

The two-mile coils of completed and proved core were wound on large drums with projecting flanges on each side, the rims of which were shod with iron tires, so that they might be rolled about as broad wheels, and made to perform their own locomotive offices as far as possible. When the core was in position on these channelled drums, the circumference of the drum was closed in carefully by a sheet of gutta percha, which thus constituted its core-filled channel a sort of cylindrical box or packing case. In this snug nest each completed coil of core was wheeled and dragged away to be transferred to the manufactory, either at Birkenhead or Greenwich.

The core-filled drums, having arrived at the factory of the Cable, the drums were mounted by axles, and kept ready so that one extremity of the length of core might be attached to the Cable as it was spun out, when the drum previously in use had been exhausted. During the unrolling of the core from the drum, it was wound tightly round by a serving of hemp, saturated with a composition made chiefly of pitch and tar, the winding being effected by revolving bobbins as the core was drawn along. This hempen serving constituted a bed for the external coat of metallic wires, and prevented the insulating sheath of gutta percha from being injured by pressure during the final stage of the construction. Each new length of core was attached to the Cable by precisely the same operation as that used at the gutta percha works in joining the two-mile coils for testing ; shortly before an old drum was exhausted, its remainder was rapidly pulled off and placed in the joiner's hands, so that it might be made continuous with the core on a new drum, before the outgoing Cable began to draw upon it.

When the core was covered in with its great coat of hemp and tar, and carefully gauged to ascertain the equality of its dimensions everywhere, it was ready to be turned into the completed Cable. This final operation was effected as the core was drawn up through the centre of a horizontally revolving wheel or table. The table turned with great rapidity, and carried near its circumference eighteen bobbins or drums. Each of these drums was filled with a strand of bright char-coal iron wire, and had two motions, one round its horizontal axis, and one round an upright pivot, inserted into the revolving table, so that it delivered its strand always towards the centre of the table as it was carried swiftly round by the revolution. The iron strand was of the same diameter as that which was used for the copper core.   There were also seven iron wires in each strand, exactly like those for the copper strand.   Eighteen iron strands were thus firmly twisted round the central core, as the " closing machine " whirled.   The core, acted on by the rollers of the machinery, rose through the middle of the table, and went up towards the ceiling.   The iron strands danced round it, as it went up, in a filmy-looking spectre-like cone, which narrowed and grew more matter-of-fact and distinct as it ascended, until it glittered in a compact metallic twist, tightly embracing the core.   The eighteen strands of seven-thread wire were used for this metallic envelope in place of eighteen simple wires of the same size as the strand, because by this means greater flexibility and strength were obtained for the weight of material employed.

Each strand machine worked day and night, and in the twenty-four hours spun ninety-eight miles of wire into fourteen miles of strand.   There were several strand machines at work in the factories, and these every twenty-four hours made 2,058 miles of wire into 294 miles of strand.   As much as thirty miles of Cable were made in a single day.   The entire length of wire, copper, and iron employed in the manufacture, amounted to 332,500 miles, enough to girdle the earth thirteen times.

As the closed Cable was completed, it was drawn out from the wall of the factory, and passed through a cistern containing pitch and tar, and was then coiled in broad pits in the outer yard (each layer of the coil having been again brushed over with pitch and tar), and there remained until embarked on board the vessel which conveyed it to its final home.   At both the Greenwich and Birken-head works, four Cables, each three hundred miles long, were simultaneously in process of construction.   These were finally united together into one continuous rope, as the Cable was stowed away in the vessel which carried it to sea.

Such is a description of the Cable finally adopted, and which when completed weighed from nineteen hundredweight to one ton per mile, and bore a direct strain of from four to five tons without breaking.

The next question which arose for consideration was, how the Cable was to be laid in the ocean.   The Great Eastern, then known as the Leviathan, alone could embrace it within her gigantic hold ; but then the vast fabric had never been tried.   She might prove a failure, and in doing so, involve that of a far greater and a far more important experiment.

It was then determined that the responsibility should be divided, and the burden be entrusted to two vessels of smaller dimensions.   The British Government placed at the service of the Company the Agamemnon line-of-battle ship, and the government of the United States of America sent over the Niagara.

D

The Agamemnon was considered to be admirably adapted for receiving the Cable, by reason of her peculiar construction ; her engines being situated near the stern, and there being amidships a magnificent hold, forty-five feet square and twenty feet deep between the lower deck and the keel. In this receptacle one half of the Cable was distributed round a central core in a compact, single, and nearly circular coil. She lay moored off the wharf at Greenwich, and the Cable was drawn into her hold by a small journeyman engine of twelve-horse power, the rope running over sheaves borne aloft upon the masts of two or three barges, so moored between the wharf and the ship as to afford intermediate support. The Niagara, though not by construction well adapted for the Cable, was rendered so by judicious alterations at Portsmouth. She arrived in the Mersey on 22nd June, and was regarded with much curiosity and interest in Liverpool, where Captain Hudson and his officers received every attention. The Cable was coiled on board her in three weeks. Cork Harbour was selected as the place where these vessels should rendezvous, and make all final arrangements; from whence they were to proceed to the completion of the task, piloted by the U.S. frigate Susquehanna and H.M. frigate Leopard, both paddle-wheel steamers of great power.

Within the barony of Iveragh, in the county of Kerry, on an island six miles long by two broad, lies the village of Knightstown and harbour of Valentia, the most westerly port in Europe. It is at the southern entrance of the open bay of Dingle towards the sea. Doulas Head on the east, and Reenadroolan Point on the west, mark the entrance to the narrows. It can boast of two forts erected by Cromwell. The Skelligs—two picturesque and rugged pinnacles of slate—pierce the surface of the sea about eight miles S.W. of the harbour ; and one of these, the " Great Skellig," crowned with a light-house, towers to a height of 700 feet.

It was decided by the Company that the Niagara should land the shore end in Valentia, and pay it out till her cargo was exhausted mid-way, where the Agamemnon was to take up the tale and carry it on to Newfoundland. The time best adapted for depositing the Cable in the ocean was determined after much thought and deliberation. The result of Lieutenant Maury's observations was, that in the months of June and July the risk of storms is very small, unless immediately on the coast of Ireland, while the records of the Meteorological Departments, both in England and America, showed that for fifty years no great storm had taken place at that period. It was finally arranged to adopt Lieutenant Maury's views, " that between the 20th July and the 10th of August both sea and air were in the most favourable condition for

R.M.Bryson,lith from a drawing by R.Dudley

London Day & Son,Limited,Lith.

VALENTIA IN 1857-1858 AT THE TIME OF THE LAYING OF THE FORMER CABLE

laying down the Cable," and that the vessels should be dispatched so as to reach the rendezvous in mid-ocean, where the Cable was to be spliced, as soon after the 20th of July as possible. It had been ascertained that the distance over which the Cable was to be laid was 1,834 miles, but 600 additional miles of Cable were provided, being an allowance of 33 per cent. of " slack."

Arrangements had been made that when the vessels joined company off Cork the entire length of the Cable should be temporarily joined up for the purpose of being tested through its entire length, as also to allow of some experiments being made to prove the efficiency of the signalling apparatus. The Cable was arranged so as to come up from the hold of the ship sweeping round a central block or core planted in the midst, to prevent any interference of the unrolling strands with one another, or too sudden turns, which might twist the Cable into kinks ; having reached the open space above the deck, it was to be wound out and in, round four grooved sheaves, geared together by cogs, and planted so firmly on girders as to render it impossible that they should be thrown out of the square. From sheaves accurately grooved the Cable proceeded three or four feet above the poop-deck, until it passed over a fifth grooved sheave standing out upon rigid arms over the stern. From this it would make its plunge into the deep still sea, and as the vessel moved away to be dragged out by its own weight, and by the hold which it would have acquired upon the bottom of the sea. The paying-out sheaves were large grooved drums, five feet in diameter, and set in a vertical plane, one directly before the other, and having a friction drum geared to them in such a way that its shaft revolved three times as fast as theirs, the axis of the drum being encircled by two blocks of hard wood, which could be gripped close upon its circumference by screw power, so as either to retard or arrest altogether the movement of the sheaves. The screw was worked by a crank, at which a trustworthy officer was stationed, to keep a wary eye upon an indicator near to express the exact amount of strain thrown upon the Cable at each instant. In the electrician's department there were to be signals every second by electrical currents passing through the entire length of the Cable, from shore-end, or from ship to ship. At the side of the vessels patent logs hung down into the water, to measure the velocity of the ship. One of these wheels, in the immersed log, was arranged to make and break an electric circuit at every revolution, a gutta percha covered wire running up from the revolving wheel on to the deck of the ship, that it might carry the current whenever the circuit was made, and record there, upon a piece of apparatus provided for the purpose, the speed of the vessel. The brakesman was to watch the tell-tale which would indicate the strain on the rope, and work his crank and loosen his grip whenever this seemed to be too great; or tighten his grip if ever the bell ceased to report that the electrical way from end to end of the Cable was free and un-impaired. An external guard had been placed over the screws of the vessels to defend the Cable from fouling in case any necessity should arise for backing the vessels. The Agamemnon had been jury-rigged for the service, her heavy masts and rigging removed, and lighter ropes and spars substituted. In the event of sudden and unforeseen storm, arrangements had been made to slip the Cable. On the decks of the paying-out vessels two large reels were placed, each wound round with two and a-half miles of a very strong auxiliary Cable composed of iron-wire only, and capable of resisting a strain of ten to twelve tons. Should the Telegraph Cable be endangered it would be divided, and the sea end attached to one of the strong supernumerary cords stored upon the reel; this being rapidly let out, would place the Cable in a depth of ocean where its safety would be secured until all danger had passed. In fine, every possible contrivance that ingenuity could devise or scientific knowledge could suggest, according to the experience then attained, had been adopted in order to secure success. Those who had toiled so long with wearied brain and anxious heart, undismayed by difficulties—not disheartened by failure, hoping when hope seemed presumptuous, but not despairing even when despair seemed wisdom, now felt that their part had been accomplished, that the means of securing the result had now passed beyond man's control, and rested solely with a Higher Power.

On the 29th of July, 1857, the U.S.N. frigate Niagara arrived at Queenstown, having been preceded by H.M.S. Leopard and H.M.S. Cyclops, which latter steamer had taken the soundings of the intended bed of the Cable. The Niagara was accompanied by the U.S.N.S. Susquehanna, to act as her convoy. H.M.S. Agamemnon had already arrived.

The Earl of Carlisle, Lord-Lieutenant of Ireland, ever anxious to give such encouragement as his presence could afford to any undertaking which promised to do good, came down from Dublin to Valentia, and attended a *déjeuner* given by the Knight of Kerry to celebrate an event in which the keenest interest was evinced, although the heart of the country was thrilled by the dreadful intelligence of Indian mutinies and revolt. The country people flocked to the little island, and expressed their joy by merrymakings, dances, and bonfires. In an eloquent speech Lord Carlisle declared that though disappointment might be in store for the promoters, it would be almost criminal to feel discouragement then—" that the pathway to great achievements has frequently to be hewn out amidst perils and difficulties, and that preliminary failure is ever the law and condition of ultimate success." These were prophetic words; in others, still to be fulfilled, " Let us hope," he said.

"We are about, either by this sun-down or by to-morrow's dawn, to establish a new material link between the Old World and the New. Moral links there have been—links of race, links of commerce, links of friendship, links of literature, links of glory; but this, our new link, instead of superseding and supplanting the old ones, is to give them a life and intensity they never had before. The link which is now to connect us, like the insect in a couplet of our poet,

'While exquisitely fine,
Feels at each thread and moves along the line.'"

If anything could overcome the tendency of men to vaticinate, it surely would be the sad history of the last few years in the United States. The condition of affairs in that lamentable period is illustrated by another passage of his lordship's speech, which also points out the inestimable value of the telegraph as a conservator of peace. "We may as we take our stand here on the extremest rocky side of our beloved Ireland, leave, as it were, behind us the wars, the strifes, and the bloodshed of the older Europe, and pledge ourselves, weak as our agency may be, imperfect as our powers may be, inadequate in strict diplomatic form as our credentials may be; yet, in the face of the unparalleled circumstances of the place and the hour, in the immediate neighbourhood of the mighty vessels whose appearance may be beautiful upon the waters, even as are the feet upon mountains of those who preach the Gospel of peace—as a homage due to that serene science which often affords higher and holier lessons of harmony and goodwill than the wayward passions of man are always apt to learn—in the face and in the strength of such circumstances, let us pledge ourselves to eternal peace between the Old World and the New. Why, gentlemen, what excuse would there be for misunderstanding? What justification could there be for war, when the disarming message, when the full explanation, when the genial and healing counsel may be wafted even across the mighty Atlantic, quicker than the sunbeam's path and the lightning's flash?" At that moment Great Britain was just disengaged from a war with Russia and a war with Persia, and was actively engaged in a war with China, and with mutinies in India. France was preparing to deal Austria a deadly blow; America looked pityingly across the Atlantic, and wondered at our folly and our crimes.

On August the 5th, 1857, the shore end of the Cable was secured in the little cove selected for the purpose in Valentia, on the cliffs above which a telegraphic station had been erected, and was hauled up amidst the greatest enthusiasm, Lord Carlisle participating in the joy and the labour.

On the evening of Friday, August 7th, the squadron sailed, and the Niagara

commenced paying out the Cable very slowly. About four miles of the shore
Cable had been payed out, when it became entangled with the machinery, by the
carelessness of one of the men in charge, and broke; all hands were engaged in
trying to underrun and join the Cable, but it was too rough, and the Niagara
came to anchor for the night. Next day a splice was made, the ship resumed
her course, and at noon on Sunday, August 9th, 95 miles had been payed out.
The paying-out gear proved to be defective in the course of the 10th. On the
evening of Tuesday, the 11th, all signals suddenly ceased. The Cable had broken
in 2000 fathoms of water, when about 330 nautical miles were laid, at a distance
of 280 miles from Valentia. At the time the ship was going from three to four
knots, and was able to pay out 5 to $5\frac{3}{4}$ miles per hour, the pressure shown by
the indicator being 3000lb., but the strain being no doubt much greater.

This loss proved fatal to the first attempt to lay the Atlantic Cable, as on
consultation among the officers and engineers it appeared to be unwise to renew
the attempt with only 1,847 miles on board the ships, or an excess of 12 per cent.
on the quantity required by the whole distance.

Nothing daunted by the failure, Mr. Field started off at once in H.M.S.
Cyclops for England, and, on his arrival, urged the immediate renewal of the
enterprise; but it was resolved by the directors in London to postpone it to
the following year. An addition to the capital of the Company was proposed and
agreed to. The greater part of the autumn was devoted to preparations for
the renewed efforts of the Company. The part of the Cable which was left was
landed at Keyham, 53 miles of the shore-end were recovered, and the Company again applied to the British and American Governments for the services
of the same vessels which had been previously lent to them. Messrs. Glass,
Elliot, & Co., were entrusted by the directors of The Atlantic Telegraph
Company to manufacture a further length of 900 miles, to replace that which
was lost or damaged, thus making a total of 3,012 miles of Cable, so as to
guard against accidents by giving an allowance of 40 per cent. of slack. The
paying-out apparatus was also improved, so that the engineer in charge alone
should control the egress of the Cable, instead of using the hand-wheel, which,
upon the former occasion, had caused much danger in rough weather.

The manufacturers of the machinery were Messrs. Easton & Amos, of South-
wark, under the superintendence of Mr. Penn, Mr. Field, Mr. Lloyd, Mr. Everett,
and Mr. Bright.

The important part of the apparatus consisted of Appold's self-regulating
brake, so adjusted and constructed as always to exert a certain amount of resistance, regulated by the revolution of the wheels to which it was applied. More

than this fixed amount of resistance, whatever it might be, it could not produce, no matter whether the machine was hot or dry, or covered with sand ; neither could it be worked at less than this amount. It was made of bars of wood laid lengthwise across the edge of the wheel, over which it lapped down firmly, and to which it was held with massive weights fixed to the ends of levers, which regulated the degree of resistance to the revolutions of the wheel, and which, of course, enabled those in charge of the machine to fix the pressure of the brake. In the new apparatus the brake was attached over two drums connected with the two main grooved wheels, round which the actual Cable passed in running out. The latter were simply broad, solid, iron wheels, each cut with four very deep grooves in which the Cable rested, to prevent it flying up or " overriding." It passed over these two main wheels, not in a double figure of eight, as in the old ponderous machine of four wheels, but simply wound over one, to and round the other, and so on four times, till it was finally payed down into the water. Thus, the wire was wound up from the hold of the vessel, passed four times over the double main wheels, connected with the brake or friction drums, past the register which indicated the rate of paying out and the strain upon the Cable, and then ran at once into the deep. The strain at which the Cable would break was 62 cwt., and to guard against any chance of mishap, not more than half this strain was put upon it. The brakes, as a rule, were fixed to give a strain of about 16 cwt., and the force required to keep the machine going, or about 8 cwt. more, was the utmost that was allowed to come upon the wire.

The brake of the paying-out machine used on the occasion of the first attempt was capable, by a movement of the hand, of exerting prodigious resistance. In the new machine any one could in a moment ease it, until there was no resistance at all beyond the 8 cwt. strain on the wire.

At a few feet from the paying-out machine, the Cable passed over a wheel, which registered precisely the strain in pounds at which the coil was running out. Facing this register was a steering wheel, similar to that of an ordinary vessel, and connected in the same way with compound levers, which acted upon the brake. The officer in charge of the apparatus stood by this wheel, and watched the register of strain or pitch of the vessel, opening the brakes by the slightest movement of his hand, and letting the Cable run freely as the stern rose. The same officer, however, could not by any possible method increase the actual strain on the Cable, which remained always according to the friction at which the brake was at first adjusted by the engineer.

All was ready for the expedition before the time indicated, and the directors and the public looked with confidence to the result. Instead of landing a shore-end at Valentia, and making a junction of the Cable, it was decided that the ships should proceed together to a point midway between Trinity Bay and Valentia, there splice the Cable, and then turn their bows east and west, and proceed to their destinations.

On Thursday, the 10th of June, 1858, H.M.S. Agamemnon and U.S.N.S. Niagara, accompanied by H.M.S. Valorous and H.M.S. Gorgon, left Plymouth, the two former having previously made an experimental cruise in the Channel with the Cables, which were very satisfactory, in all respects.

Experienced mariners gazed with apprehension at their depth in water as they left the shore. It was, however, such glorious weather as to cause some anxiety lest there should be no wind, and that the stock of coals might be exhausted before their mission was accomplished. Before midnight, however, a gradually increasing gale gathered to a storm, while the barometer marked 29°. For seven consecutive days the tempest, so eloquently described by Mr. Woods in the *Times*, continued, the Agamemnon under close-reefed topsails striving to reach the rendezvous, Lat. 52° 2′, Long. 33° 18′, rolling 45 degrees, and labouring fearfully.

On the 19th and 20th the gale reached its height. The position of the ship, carrying 2,840 tons of dead-weight, badly stowed, had become most critical, from her violent lurching as she sunk into the troughs of the sea, and struggled violently to right herself—the coal bunkers gave way, and caused alarm and confusion. Were the masts to yield, the ship would rock still more violently, the Cable would shift, and carry every one with it to destruction. Captain Preedy had but two courses open in order to save the ship without sacrificing the Cable—either was fraught with peril—to wear the ship, or to run before the gale and risk the chances of being pooped by the monster seas in pursuit.

On the 21st the Agamemnon was enabled to bear up for the rendezvous in mid-ocean, which she reached on the 25th, after sixteen days of danger and apprehension, her companion, the Niagara, having passed through the dreadful ordeal with less danger and difficulty.

At half-past two o'clock on the 26th, the Agamemnon and Niagara first spliced the Cable; it however became foul of the scraper on the latter ship, and broke. A second splice was immediately made, and the vessels started. The Agamemnon had paid out 37½ miles, when suddenly the continuity of the electric current ceased, and the electricians declared that the Cable had broken at the bottom. As the Niagara was hauling in the Cable, of which she had payed out 43 miles, it snapped close to the ship.

On the 28th, the third and final splice was effected. The Niagara started N.W. ¾ N. At 4 p.m. on the 29th, when 111 miles had been paid out, the electricians on board reported that continuity had ceased. The cause was soon known. The Agamemnon had run 118 miles, and paid out 146 miles of Cable, when the upper deck coil became exhausted. Speed was slackened, in order to shift the Cable to the lower deck, when suddenly it snapped, without any perceptible cause, under a strain of only 2200 pounds. The weather was calm; the speed moderate—about five knots; the strain one-third less than breaking strain; everything favourable; and yet the Cable parted, silently and suddenly. The Niagara had to cut the Cable, as she had no means of recovering the portion payed out, and lost 144 miles of it.

On the 12th July, the Agamemnon, after an eventful cruise of thirty-three days, reached Queenstown, having left the rendezvous on the 6th, whither she had gone in the hope of meeting the Niagara. A special meeting of the Company was called, and the expedition was ordered to go to sea. There was still quite sufficient Cable remaining, and it was determined to make another attempt immediately. The way in which the Cable parted on the third occasion was the only thing calculated to create doubt and apprehension. The two other breakages might be accounted for, and guarded against for the future, but there was something in the latter not so easy of explanation, and which seemed to point to some mysterious agency existing in the depths of the ocean, beyond the perception of science or man's control.

At midnight on the 28th of July, 1858, the Agamemnon and Niagara once more met in mid-ocean, and on the following morning spliced the Cable, which was this time destined to tend so much towards solving the great problem. On the 30th, 265 miles had been paid out. On the 31st, 540 miles. On the 1st August, 884 miles. On the 2nd, 1256 miles. On the 4th, 1854 miles; and on the 5th, 2022 miles. The Agamemnon now anchored in Dowlas Bay, Valentia, and preparations were made to join the ocean and shore ends. On the same day, at 1·45 a.m., the Niagara anchored in Trinity Bay, Newfoundland, and in an hour after she received a signal across the Atlantic that the Cable had been landed from the Agamemnon.

Mr. Field at once telegraphed the news to the New York press, and the intelligence flew all over the Union, where it was received with the most extraordinary manifestations of delight. The information was received more equably in England.

On the 7th of August, many an anxious heart was lightened by reading in the *Times* the following telegram :—

"VALENTIA, *August 6th*.

" End of Cable safely landed, close by pier, at Knightstown, being carried on the paddle-boxes of the Valorous—expect to be open to public in three weeks."

Mr. Field's dispatch to the Associated Press of New York was followed by two to the President, to which Mr. Buchanan sent a suitable reply. A message was sent to the Mayor of New York also, to which an answer was returned next day.

On August the 9th the telegraphic wires reported that " Newfoundland still answered, but only voltaic currents."

On the 10th it was stated "Coil currents had been received—40 per minute easily"—followed by the modest words, " Please send slower for the present."

On the 14th a message of 14 words was transmitted, and on the 18th the Directors in England thus spoke to their brethren in the other hemisphere : "Europe and America are united by telegraphic communication. 'Glory to God in the highest, on earth peace, goodwill towards men.'" This message occupied 35 minutes in transmission. It was rapidly followed by a message from the Queen of England to the President of America, which occupied 67 minutes in transmission, and was repeated. The text was as follows :—

" To THE PRESIDENT OF THE UNITED STATES, WASHINGTON :

" The Queen desires to congratulate the President upon the successful completion of this great international work, in which the Queen has taken the deepest interest.

" The Queen is convinced that the President will join with her in fervently hoping that the Electric Cable which now connects Great Britain with the United States will prove an additional link between the nations whose friendship is founded upon their common interest and reciprocal esteem.

" The Queen has much pleasure in communicating with the President, and renewing to him her wishes for the prosperity of the United States."

THE REPLY OF THE PRESIDENT.

" *Washington City, August* 16, 1856.

"To HER MAJESTY VICTORIA, QUEEN OF GREAT BRITAIN :

" The President cordially reciprocates the congratulations of Her Majesty the Queen on the success of the great international enterprise accomplished by the science, skill, and indomitable energy of the two countries. It is a triumph

R.M.Bryson, lith from a drawing by R.Dudley.

London, Day & Son, Limited, Lith.

TRINITY BAY, NEWFOUNDLAND. EXTERIOR VIEW OF TELEGRAPH HOUSE IN 1857-1858

more glorious, because far more useful to mankind, than was ever won by conqueror on the field of battle.

"May the Atlantic Telegraph, under the blessing of Heaven, prove to be a bond of perpetual peace and friendship between the kindred nations, and an instrument destined by Divine Providence to diffuse religion, civilisation, liberty, and law throughout the world. In this view will not all nations of Christendom spontaneously unite in the declaration that it shall be for ever neutral, and that its communications shall be held sacred in passing to their places of destination, even in the midst of hostilities?

(Signed)   "JAMES BUCHANAN."

On the same day a message was received from Mr. C. Field, consisting of 38 words, which occupied 22 minutes in transmission.

The mighty agency which had been made subservient to the dictates of man, had touched the hearts of two nations by expressing mutual esteem and respect, but had not yet exercised its higher prerogatives. On the 21st of August it flashed tidings of great joy, and brought relief to those who, but for it, would have languished in very weariness and pining. The Europa and Arabia, each thickly freighted with human lives, had come into collision in mid-ocean. So much was known, but there was nothing to appease the anxiety of those whose friends and relatives were on board. Fourteen days must elapse before the arrival of the next steamer. Within fourteen hours, however, the Atlantic telegraph wires allayed intense dread and anxious fears : " Newfoundland.—Europa and Arabia have been in collision—one has put into St. John's—no lives are lost—all well."

On the 25th of August it was announced that "the Cable works splendidly," and shortly after the New York journals recorded how the entire continent had gone mad for very joy, how feasting was the order of the day, and how American intellect had achieved the greatest scientific triumph of the age.

On the 7th of September, 1858, the following letter appeared in the *Times*, addressed to the editor :—

"*September 6th,* 1858.

"SIR,—I am instructed by the Directors to inform you that, owing to some cause not at present ascertained, but believed to arise from a fault existing in the Cable at a point hitherto undiscovered, there have been no intelligible signals from Newfoundland since one o'clock on Friday the 3rd inst. The Directors are now in Valentia, and, aided by various scientific and practical electricians, are investigating the cause of the stoppage, with a view to remedying the existing difficulty. Under these circumstances no time can be named at present for opening the wire to the public.

"GEO. SAWARD."

Such was the foreshadowing of the great calamity that was so soon to follow. Public excitement became intense. The market value of the Atlantic Telegraph Stock assumed a downward tendency, and fell rapidly. But the projectors had not been idle. A rigid inquiry had been immediately instituted by Professor Thomson, Mr. Varley, and Sir Charles Bright, which enabled them to arrive at a conclusion that the fault must lie on the Irish coast. Consequently the Cable was underrun for three miles, cut and tested; but no defect being found, it was again spliced. During all this period its electrical condition had become so much deteriorated that such messages as passed required to be constantly repeated.

So matters went, hope and fear alternating, until the insulation of the wire became suddenly worse, and at last the signals ceased to be intelligible at Newfoundland altogether. Scientific inquiry tended to show that the fault lay about 270 miles from Valentia, at the mountain range which divides the depths of the Atlantic from the shallow water on the Irish shore. This steep range, or sloping bank, which, on being sounded, showed a difference of 7,200 feet in elevation in a distance of eight miles, had been crossed by the Agamemnon an hour before the expected time, and it was said a sufficient quantity of slack had not been thrown out, so that the Cable was suffered to hang suspended in the water. But this was of course mere conjecture, and the failure most probably was precipitated by injudicious attempts to overcome defective insulation by increased battery power.

The conclusions finally arrived at by the Scientific Committee appointed to report as to the causes of the failure of the Cable were, first, that it had been manufactured too hastily; secondly, that a great and unequal strain was brought on it by the machinery; and thirdly, that the repeated coilings and uncoilings it underwent served to injure it. To such causes was the failure to be attributed, not to any original defect in the gutta percha.

Mr. Varley stated his opinion that there must have been a fault in the Cable while on board the Agamemnon, and before it was submerged; but none of the theories accounted for the destruction of a Cable on which half a million of money had been expended, and which (if successful) two governments had contracted to subsidise to the gross amount of 28,000l. yearly. Thus were annihilated, silently

G. McCulloch, lith. from a drawing by R. Dudley.

TELEGRAPH HOUSE. TRINITY BAY. NEWFOUNDLAND. INTERIOR OF "MESS ROOM" 1858.

London. Day & Son, Limited, Lith.

and mysteriously, all those hopes which had survived so many disappointments, and which for a moment had been so abundantly realised.

But in England, as no ebullitions of joy had been indulged in when success seemed certain, neither was there now any yielding to despair.

In the month of April, 1860, the Directors of the Atlantic Telegraph Company sent out Captain Kell and Mr. Varley to Newfoundland to endeavour to recover some portion of the Cable; their efforts showed that the survey which had been taken must have been very insufficient, and the ground was much worse than was expected. They recovered five miles of the Cable, and ascertained two facts, namely, that the gutta percha was in no degree deteriorated, and that the electrical condition of the core had been improved by three years' submersion. In 1862 several attempts were also made to recover some of the Cable from the Irish side, but with no practical advantage; and in consequence of violent storms the attempt was abandoned.

The great Civil War in America stimulated capitalists to renew the attempt; the public mind became alive to the importance of the project, and to the increased facilities which promised a successful issue. Mr. Field, who compassed land and sea incessantly, pressed his friends on both sides of the Atlantic for aid, and agitated the question in London and New York.

On the 20th of December, 1862, the Atlantic Company issued its prospectus, setting forth the valuable privileges it had acquired—amongst others, the exclusive right to land telegraph wires on the Atlantic coast of Labrador, Newfoundland, Prince Edward's Island, and the State of Maine—and invited public subscriptions. The firm of Glass, Elliot, & Co., sent in tenders to provide a Cable at a cost of £700,000; a sum of £137,000, being 20 per cent. upon the capital of the Company, to be paid to them in old unguaranteed shares of the Company, provided they were successful.

On the 4th of March, 1863, a large number of the leading merchants in New York assembled in the Chamber of Commerce in that city, for the purpose of hearing some new and interesting facts relative to the Atlantic Telegraph enterprise. The many advantages which would arise to America were apparent, and, among others, was the improvement of the agricultural position of the country by extending to it the facilities, already enjoyed by England and France, of commanding the foreign grain markets; as well as the avoidance of misunderstandings between America and other countries.*

---

* Short-lived as was the former Cable, it had survived long enough to prove its value in a financial point of view. Amongst 400 messages which it had transmitted, was one that had been dispatched from London in the morning and reached Halifax the same day, directing "that the 62nd Regiment were not o return to England." This timely warning saved the country an expenditure of 50,000*l.*

Since 1858, what was a mere experiment had become a practical reality. The Gutta Percha Company had prepared no less than forty-four submarine Cables, enclosing 9000 miles of conducting wire, which were in daily use, and not one of which had required to be repaired, except at the shore end, where they were exposed to ships' anchors. At the meeting in New York, Mr. Field read a letter from Glass, Elliot, & Co., in which they offered to undertake to lay the Cable between Ireland and Newfoundland on the most liberal conditions. The terms which they proposed were,—First, that all actual disbursements for work and material should be recouped each week : secondly, that when the Cable was in full working order, 20 per cent. on the actual profits of the Company should be paid in shares to be delivered monthly, while at the same time they offered to subscribe £25,000 towards the ordinary capital of the Company. The English Government also agreed to guarantee interest on the capital at 8 per cent., during the operation and working of the Cable, and to grant a yearly subsidy of £14,000. Mr. Field further directed the attention of the meeting to the line to San Francisco (a single State), as evidence of what business might be expected. The estimated power of the Cable was a minimum of 12 and a maximum of 18 words per minute. If it were to be worked for sixteen hours per day for 300 days in each year, at a charge of 2s. 6d. per word, the income would amount to £413,000 a year, which would be a return of 40 per cent. upon a single Cable. After the failure of the last Cable a Commission of Inquiry, consisting of nine members, had sat for two years, and, by their report, afforded valuable information. The British Government had also dispatched surveying expeditions, which reported most favourably as to Newfoundland. In reference to the objection, that in case of war the Cable would be under the sole control of the English Government, it was to be remembered that it would be laid under treaty stipulations.

After a lengthened discussion on various matters connected with the project. it was proposed by Mr. A. Low, and unanimously resolved, "That, in the opinion of this meeting, a Cable can, in the present state of telegraphic science, be laid between Newfoundland and Ireland with almost absolute certainty of success, and will when laid prove the greatest benefit to the people of the two hemispheres, and also profitable to the shareholders. It is, therefore, recommended to the public to aid the undertaking."

Messrs. Glass, Elliot, & Co. had long successfully manufactured Cables in accordance with all the improvements that had taken place in machinery, as well as in the manufacture of gutta percha, since the laying of the Cable of 1858. Their experience as contractors in laying lines might be estimated by

R.M.Bryson, lith. from a drawing by R.Dudley.

London. Day & Son, Limited, lith.

H. M. S. "AGAMEMNON" LAYING THE ATLANTIC TELEGRAPH CABLE IN 1858.-A WHALE CROSSES THE LINE.

the report of the Jurors of the Exhibition of 1862. They had been identified with the history of submarine telegraphy from its earliest existence, and now, having previously incorporated the Gutta Percha Company, they accepted the offer made by capitalists of influence and became absorbed in "The Telegraph Construction and Maintenance Company," of which Mr. Pender, M.P., was chairman, and Mr. Glass managing director.

The British Government were willing to assist by subsidy and guarantee, and there lay the Great Eastern, the only vessel in the world suited for the undertaking, seeking for a purchaser. The huge ship, which cost £640,000, was chartered by the Directors of the Telegraph Construction and Maintenance Company, who seemed bent upon solving the problem of its existence, and on showing what great things it was destined to accomplish. Captain James Anderson, an accomplished officer of the Cunard line, was asked to take the command, and received leave to do so, and it was with satisfaction the Directors learned his willingness to undertake the task.

In May, 1864, a contract previously entered into was ratified, providing that all profit should be contingent on success, and that all payments were to be made in unissued shares of the Atlantic Telegraph Company. A resolution was also passed, authorising the raising of additional capital by the issue of 8 per cent. guaranteed shares, of which Glass, Elliot, & Co., were to receive 250,000l., and also 100,000l. in debentures. The form of the Cable selected was similar in its component parts to that of 1858, but widely different in the construction and quality of the materials. It had been reported on most favourably by the Committee of Selection, and was at once accepted by the contractors; the Directors of the Company recognising the assiduity and skill of Mr. Glass in the investigations as to the best description of Cable.

The following official account * states so minutely every particular connected with the Cable during the process of formation, down to its shipment on board the Great Eastern, that no better description can be given :—

It differed from the Cable of 1857-8, as to its size, as to the weight and method of application of the materials of which it was composed, as to its specific gravity, and as to the mode adopted for its external protection.

For the same reason as before, the copper conductor employed in the Cable was not a solid rod, but a strand, composed of seven wires, each of which gauged 048 parts of an inch. It was found practically that this form of conductor, in which six of the wires were laid in a spiral direction around the seventh, was a

* Communicated to the *Mechanics' Magazine.*

most effectual protection against the sudden or complete severance of the copper wire.

The severance, or "breach of continuity," as it is usually called, is one of the most serious accidents that can happen to a submerged Cable, when unaccompanied by loss of insulation—owing to the great difficulty in discovering the locality of such a fault. Even the best description of copper wire can seldom be relied upon for equality of strength throughout, and in some instances an inch or even a less portion of the wire will prove to be slightly crystallised, and consequently incapable of resisting the effects of coiling or paying out if brought to bear upon the part, though no external difference be at all apparent between the weak portion and the remainder of the sample. By proceeding, however, as in the present case, the conductor was divided into seven sections, and the risk of seven weak places occurring in the same spot being exceedingly remote, the probability of a breach of continuity in a strand conductor was almost *nil*.

The weight of the new conductor was nearly three times that of the former one —being 300 pounds to the nautical mile against 107 pounds per knot to the conductor of 1857. The adoption of this increased weight had reference to the increase of commercial speed in the working of the new Cable expected to accrue therefrom, and was founded upon the principles of conduction and induction, now well understood, which consist in the law that the conductivity of the conductor is as its sectional area, while its inductive capacity (whereby speed of transmission is retarded) is as its circumference only ; and, as the maximum speed at which the original Cable was ever worked did not exceed two and a-half words per minute, it would follow by calculation, taking into account the thickness of the dielectric surrounding the present conductor, that, using the same instruments as in 1858, a speed of three and a-half to four words per minute might be expected from the new Cable ; but it was stated by the electricians that owing to the improved modes of working long Cables that have been discovered since 1858, an increase of speed up to six or even more words per minute might be secured by the adoption of suitable apparatus.

The purity of the copper employed, a very important item, affecting the rate of transmission, had been carefully provided for. Every portion of the conductor was submitted to a searching test, and all copper of a lower conductivity than 85 per cent. of that of pure copper was carefully rejected.

The covering of the conductor with its dielectric or insulating sheath was effected as follows :—The centre wire of the copper strand was first covered with a coating of gutta percha, reduced to a viscid state with Stockholm tar, this being the preparation known as "Chatterton's Compound." This coating must be so thick that, when the other six wires forming the strand were laid spirally and tightly round it, every interstice was completely filled up and all air excluded. The object of this process was two-fold; first, to prevent any space for air between the conductor and insulator, and thus exclude the increase of inductive action attendant upon the absence of a perfect union of those two agents, and, second, to secure mechanical solidity to the entire core; the conductors of some earlier Cables having been found to be to some extent loose within the gutta percha tube surrounding them, and thereby much more liable to permanent extension, mechanical injury, and imperfect centricity than those in which the preliminary precaution just described had been made use of. The whole conductor next received a coating of Chatterton's Compound outside of it; this, when the core was completed, quickly solidified, and became almost as hard as the remainder of the subsequent insulation. It was then surrounded with a first coating of the purest gutta percha, which being pressed around it while in a plastic state by means of a very accurate die, formed a first continuous tube along the whole conductor. Over this tube was laid by the same process a thin covering of Chatterton's Compound, for the purpose of effectually closing up any possible pores or minute flaws that might have escaped detection in the first gutta percha tube. To this covering of Chatterton's Compound succeeded a second tube of pure gutta percha, then another coating of the compound, and so on alternately until the conductor had received in all four coatings of compound and four of gutta percha. The total weight of insulating material thus applied was 400 pounds to the nautical mile, against 261 pounds in the Cable of 1857-8.

The core, completed as described, and which had previously and repeatedly been under electrical examination, was at length submerged in water of a temperature of 75 deg. Fah., and so remained during twenty-four hours. This was done that the subsequent electrical tests for conductivity and insulation might be made under circumstances the most unfavourable to the manufacture, from the well-known fact, that the insulating power of gutta percha is sensibly decreased by heat. It also ensures uniformity of condition to the core under test, and, the temperature in which it was tested being higher by 20 deg. than that of the water of the North Atlantic, there was plenty of margin against any disappointment from the effects of temperature after submersion. At the expiration of the term of soaking, the coils of core submitted to that process were expected to show an insulation of not less than 5,700,000 of Varley's standard units, or of 150,000,000 of those of Siemens's standard. This of itself was a very severe test, but no portion of the core showed a less perfection than that of double of either of the above high standards.

Having passed this ordeal, and having been tested on separate instruments

F

and by a different electrical process by the officers of the Atlantic Telegraph Company, in order to verify the observations of the contractors, the core was tested for insulation under hydraulic pressure, after which it was carefully unwound from the reels on which it had been wound for that purpose, and every portion was carefully examined by hand as it was rewound on to the large drums on which it was sent forward to the covering works at East Greenwich, to receive its external protecting sheath. It was then again submerged in water, and required once more to pass the full electrical tests above referred to. Finally, each reel of core was very carefully secured and protected from injury, and in this state was sent to East Greenwich, where it was immediately placed in tanks provided for it. In these it was covered with water, and the lids of the tanks being fastened down and locked, it remained until demanded for completion.

The manufacture and testing of the "core" of the Atlantic Cable having been completed at the Gutta Percha works as described, a telegraphic line was thereby produced which, without further addition of material or substance, beyond that of copper and gutta percha, proportionable to any required increase in its length, would be perfect as an electrical communicator through the longest distances and in the deepest water, in which element moreover it appears to be chemically indestructible, if the experience of some fourteen years may be taken as evidence. At this point, however, the final form to be assumed by the deep-sea Cable was subject to important mechanical considerations, which came into play across the path of those purely electrical; and upon the manner in which these considerations are met and dealt with, depend, not merely the primarily successful submersion, but the ultimate durability and commercial value of deep-sea Cables.

The problem in the case of the Atlantic Telegraph enterprise may be thus stated. Given a submarine telegraph core like that already described, constructed on the best known principles and perfect as to its electrical conductivity and insulation—it is required to lower the same through the sea to a maximum depth of two and a-half miles, so as not merely not to allow the insulating medium to be torn or strained, but so as not even to bring its normal elasticity into play against the more tensile but perfectly inelastic material of the conductor. For if the core were lowered into very deep water like that referred to without further protection, even supposing it to escape actual fracture by the adoption of extraordinary precaution and by the aid of fine weather, it is evident that whenever, as would be highly probable, either in the act of paying out, during the lifting or manœuvring of the ship, or even from the effects of its own weight, the gutta percha sheath became extended to the limit of its elasticity, the copper in the centre would be stretched to a corresponding extent, and, the tension being removed, the gutta percha in returning to its original length would pull back the now elongated copper, which thenceforward would in every such case " buckle up," and exert a constant lateral thrust against the gutta percha; ending, probably, in its ultimate escape to the outside, and the consequent destruction of the core as an electrical agent. Moreover, in the event of an electrical fault being discovered in any submerged portion of the Cable during the process of " paying-out" in deep water, it is of paramount importance towards its recovery and repair, that the engineer should have such an assurance in the quality and strength of his materials as will enable him confidently to exert a known force in hauling back the injured part, without apprehension of damage to the vital portion of the Cable.

The solution of this question must therefore be found in adding mechanical strength externally to the core, by surrounding it with such materials and in such a manner as to relieve it from all that strain which it will unavoidably meet in depositing it in its required position. In the case of the original Atlantic Cable this was attempted by first surrounding the core with tarred hemp, which in its turn was enveloped spirally by eighteen strands of iron wire; each strand consisting of seven No. 22½ gauge wires. The entire weight of the Cable so formed was, in air 20 cwt. per knot, and in water 13·3 per knot. Being capable of bearing its own weight in about five miles perpendicular depth of water, and the greatest depth on the route being two-and a half miles, its strength was calculated at about as much again as was absolutely requisite for the work. This was thought at the time to be a sufficient margin, and certainly in 1858, owing to the greatly improved machinery employed, this Cable was payed-out with great facility and without undue strain, although portions of it had been lost by breaking during several previous attempts in the same summer. Subsequent investigation and experience, however, led to the conclusion, that in respect, not only to its mechanical properties, but especially with regard to its relative specific gravity, and to other points in its construction, the Cable of 1858 was very imperfect; and, with a view to ensure every practicable improvement in the structure of their new line, the promoters of the undertaking, so soon as they found themselves in funds, during 1863, issued advertisements with a view to stimulate inquiry into the subject, inviting tenders for Cables suitable for the proposed work. The specimens that were sent in, as the result of this public appeal, were submitted to the scientific advisers of the Company, who, after careful experiments with all the specimens, unanimously recommended the Atlantic Company to adopt the principle of the Cable proposed by Glass,

Elliot, & Co., whose experience and success in this description of work are well known. The Committee, however, stipulated that they should settle the actual material of which the Cable was to be ultimately composed, and that these should be carefully and separately experimented on before finally deciding upon it; and in consequence of this stipulation upwards of one hundred and twenty different specimens, being chiefly variations of the principle adopted by the Committee, were manufactured and subjected to very severe experiment, as were also the various descriptions and quantities of iron, hemp, and Manilla proposed as components of these respective Cables. The result of it all was that the Committee recommended the Cable that was adopted as being, in their opinion, " the one most calculated to insure success in the present state of our experimental knowledge respecting deep-sea Cables," taking care at the same time, by enforcing a stringent specification and constant supervision, to guard against any possible laxity in the details of its construction. The Cable so decided on weighed $35\frac{3}{4}$ cwt. per knot in air, but in water it did not exceed 14 cwt., being only a fraction heavier in that medium than the old Cable, though bearing more than twice the strain—the breaking strain of the new Cable being 7 tons 15 cwt. In water it was capable of bearing eleven miles of its own length perpendicularly suspended, and consequently had a margin of strength of more than four and a-half times that which was absolutely requisite for the deepest water. The core having been received from the gutta percha works, and carefully tested to note its electrical condition, was first taken to receive its padding of jute yarn, whereby the gutta percha would be protected against any pressure from the external iron sheath, which latter succeeded the jute. On former occasions this padding of jute had been saturated in a mixture of tar before being applied to the gutta percha; but experience had shown that this proceeding might lead to serious fallacies as to the electrical state of the core, cases having been repeatedly found where faults existed in the core itself—amounting to an almost total loss of insulation—which, however, were only discovered after being submerged and worked through, owing to the partial insulation conferred for a time upon the bad place by means of the tarred wrapping. The Atlantic core, therefore, was wrapped with jute which had been simply tanned in a solution of catechu, in order to preserve it from decay, and as fast as the wrapping proceeded the wrapped core was coiled into water, in which, not only at this stage, but ever afterwards until finally deposited in the sea, the Cable, complete or incomplete, was stored, and the water being able to freely pass through the tarred jute to the core, the least loss of insulation was at once apparent by the facility offered by the water to conduct away to earth the whole or a portion of the testing current.

R.M. Bryson, lith. from a drawing by R. Dudley.

London, Day & Son, Limited, Lith

COILING THE CABLE IN THE LARGE TANKS AT THE WORKS AT GREENWICH.

The iron wire with which the jute cover was surrounded was specially prepared for this purpose, and is termed by the makers (Messrs. Webster & Horsfall) "Homogeneous Iron." It was manufactured and rolled into rods at their works at Killamarsh, near Sheffield, and drawn at their wire factory at Hay mills, near Birmingham. This wire approaches to steel in regard to strength, but by some peculiarity in the mode of preparing it, is deprived entirely of that springiness which prohibits altogether the use of steel as a covering for the outsides of submarine cables. Ten wires were laid spirally round the core, and each of these wires was of No. 13 gauge, and was under contract to bear a strain of 850 to 1,100 lb., with an elongation of half an inch in every fifty inches within those breaking limits. The Cable, as completed and surrounded by these wires, had not the slightest tendency to spring, as would be the case if the metal were hard steel, and could be handled with great facility.

Before, however, these ten wires surrounded the core, each separate wire had to be itself covered with a jacket of tarred Manilla yarn, the object of which is at once to protect the iron from rust and to lighten the specific gravity of the mass, while adding also in some degree to the strength of the external portion of the Cable. The wire was drawn horizontally forward over a drum through a hollow cylinder, on the outside of which bobbins filled with Manilla yarn revolved vertically, and the yarns from these bobbins, being made to converge around the wire as it issued from the end of the cylinder, were thus spun tightly round the former. These Manilla-covered wires being wound upon large drums ready for use, the core, which we left some time back surrounded with jute, was passed round several sheaves, which conducted it below the floor of the factory, from whence it was drawn up again through a hole in the centre of a circular table, around the circumference of which were ten receptacles for ten drums, containing the Manilla-covered wire. Between these drums massive iron rods, fastened to the circumference of the table, rose, and converged around a small hollow cone of iron through the upper flooring of the factory, at a height of 12 or 14 feet above the table. The jute-covered core was pulled up vertically, and passed on straight through the hollow interior of the cone already mentioned, which latter formed the apex of the converging rods. This done, the ten wires from the ten drums were drawn up over the outside of the same cone, and as they rose beyond it converged around the core, which latter, being free from the revolving part of the machinery, was simply drawn out; while the circular table being now set revolving by steam power, the ten wires from the drums were spun in a spiral around the core, thus completing the Cable, which was hauled out of the factory by the hands of men, who at the same time coiled it into large iron tanks, where it was covered with water, and was daily subjected to the most careful electrical tests, both by the very experienced staff of the contractors and by the agents of the Atlantic Telegraph Company.

The distance from the western coast of Ireland to the spot in Trinity Bay, Newfoundland, selected as the landing-place for the Cable, was a little over 1,600 nautical miles, and the length of Cable contracted for, to cover this distance, including the "slack," was 2,300 knots, which left a margin of 700 knots, to cover the inequalities of the sea-bed, and to allow for contingencies. On the first occasion 2,500 statute miles were taken to sea, the distance to the Newfoundland terminus on that occasion being 1,640 nautical miles; and, after losing 385 miles in 1857, and setting apart a further quantity for experiments upon paying-out machinery, sufficient new Cable was manufactured to enable the Niagara and Agamemnon to sail in 1858 with an aggregate of 2,963 statute miles on board the two ships, of which about 450 statute miles were lost in the two first attempts of that year, and 2,110 miles were finally laid and worked through.

The greatly increased weight and size of the Cable would have made the question of stowage a very embarrassing one had it not been for the existence of the Great Eastern steamship, there being no two ordinary ships afloat that would be capable of containing, in a form convenient for paying-out, the great bulk presented by 2,300 miles of a Cable of such dimensions. This bulk, and the now acknowledged necessity for keeping Cables continuously in water, made their influence to be felt in a very expensive manner to the Company and to the contractors throughout the progress of the work, even at this early stage. The works at Morden Wharf had to be to a very large extent remodelled to meet these contingencies. Eight enormous tanks, made of five-eighths and half-inch plate iron, perfectly watertight, and very fine specimens of this description of work, were erected on those premises, and these tanks then received an aggregate of 80 miles of Cable per week. Four of the tanks were circular in shape, and each contained 153 miles of cable, being 34 ft. in diameter and 12 ft. deep. The other four were slightly elliptical, being 36 ft. long by 27 ft. wide, and 12 ft. deep, and contained each 140 miles. The contents of all these, as they became full, were transferred to the Great Eastern at Sheerness, for which service the Lords of the Admiralty granted the loan of two sailing-ships, laid up in ordinary at Chatham, namely—the Amethyst and the Iris.* These ships had to undergo very considerable alteration to

---

* It may here be stated that Admiral Talbot, in command at the Nore, gave every aid to the undertaking; and that Captain Hall, of the Sheerness Dockyard, was indefatigable and most serviceable in forwarding the work whilst the Great Eastern lay in the Medway and at the Nore.

F. Jones. lith. from a drawing by R. Dudley.

London, Day & Son, Limited, Lith

THE CABLE PASSED FROM THE WORKS INTO THE HULK LYING IN THE THAMES AT GREENWICH.

T. Picken, lith. from a drawing by R. Dudley

London, Day & Son, Limited, Lith

THE OLD FRIGATE WITH HER FREIGHT OF CABLE ALONGSIDE THE "GREAT EASTERN" AT SHEERNESS

render them suitable for the work, portions of the main deck having to be removed—fore and aft—to make room for watertight tanks, which here, as elsewhere, were to be the medium for holding the Cable. The dimensions of the two tanks on board the Amethyst were 29 ft. diameter by 14 ft. 6 in. in depth, and each held 153 miles of Cable ; of those on the Iris, one was 29 ft. diameter and 14 ft. 6 in. deep, and held 153 miles, and the other held 110 miles, and was 24 ft. wide, and 17 ft. deep.

The Great Eastern steamship was fitted up with three tanks to receive the Cable, one situated in the forehold, one in the afterhold, and the third nearly amidships. The bottoms and the first tier of plates were of five-eighths iron, and each tank, when completed to this height, and tested as to its tightness by filling it with water, and found or made to be perfectly watertight, was let down from its temporary supports on to a bed of Portland cement, three inches in thickness, and the building up and riveting of the remaining tiers was continued. The beams beneath each tank were shored up from the floor beneath it down to the kelson with nine inches Baltic baulk timber, and it will give some idea of the magnitude of the work to state that upwards of 300 loads of this material were required for this purpose alone. The dimensions of the fore tank were 51 ft. 6 in. diameter by 20 ft. 6 in. in depth, and its capacity was for 693 miles of Cable. The middle tank was 58 ft. 6 in. broad, and 20 ft. 6 in. deep, and held 899 miles of Cable, and the after tank was 58 ft. wide and 20 ft. 6 in. deep, and contained 898 miles. The three tanks were therefore capable of containing in all 2,490 miles of the new Cable.

The experience gained on board the Agamemnon and Niagara, and the practical knowledge obtained by the telegraphic engineers, were turned to good account in erecting the new machinery on the deck of the Great Eastern for paying-out the Cable.

Over the hold was a light wrought-iron V wheel, the speed of which was regulated by a friction wheel on the same shaft. This was connected with the paying-out machinery by a wrought-iron trough, in which, at intervals, were smaller wrought-iron V wheels, and at the angles vertical guide wheels. The paying-out machinery consisted of a series of V wheels and jockey or riding wheels (six in number); upon the shafts of the V wheels were friction wheels, with brake straps weighted by levers and running in tanks filled with water : and upon the shafts of the jockey wheels were also friction straps and levers, with weights to hold the Cable and keep it taut round the drum. Immediately before the drum was a small guide wheel, placed under an apparatus called the knife, for keeping the first turn of the Cable on the drum from riding or getting over another turn. The knives, of which there were two, could be removed and adjusted with the greatest ease by slides similar to a slide-rest of an ordinary turning-lathe. One knife only was used, the other being kept ready to replace it if necessary. The drum, round which the Cable passed, was 6 feet diameter and 1 foot broad, and upon the same shaft were fixed two Appold's brakes, running in tanks filled with water. There was also a duplicate drum and pair of Appold's brakes fitted in position and ready for use in case of accident. Upon the overhanging ends of the shafts of the drums driving pulleys were fitted, which could be connected by a leather belt for the purpose of bringing into use the duplicate brakes, if the working brakes should be out of order. Between the duplicate drum and the stern wheel were placed the dynamometer and intermediate wheels for indicating the strain upon the Cable. The dynamometer wheel was placed midway between the two intermediate wheels, and the strain was indicated by the rising or falling of the dynamometer wheel on a graduated scale of cwts. attached to the guide rods of the dynamometer slide. The stern wheel, over which the Cable passed when leaving the ship, was a strong V wheel, supported on wrought-iron girders overhanging the stern, and the Cable was protected from injury by the flanges of this wheel by a bell-mouthed cast-iron shield surrounding half its circumference.

Close to the dynamometer was placed an apparatus similar to a double-purchase crab, or winch, fitted with two steering wheels, for lifting the jockey or riding wheels with their weights and the weights on the main brakes of the drum, as indications were shown upon the dynamometer scale.

All the brake wheels ran in tanks supplied with water by pipes from the paddle-box tanks of the ship.

The Cable passed over the wrought-iron V wheel over the tank along the trough, between the V wheels and jockey wheels in a straight line ; four turns round the drum where the knife comes into action over the first intermediate wheel, under the dynamometer wheel, and over the other intermediate and stern wheels into the sea.

This dynamometer was only a heavy wheel resting on the rope, but fixed in an upright frame, which allowed it to slide freely up and down, and on this frame were marked the figures which showed exactly the strain in pounds on the Cable. Thus, when the strain was low the Cable slackened, and the dynamometer sunk low with it ; when, on the contrary, the strain was great, the Cable was drawn " taut," and on it the dynamometer rose to its full height. When it sunk too low, the Cable was generally running away too fast, and the brakes had to be applied to check it ; when, on the contrary, it rose rapidly the tension was dangerous, and the brakes had to be almost opened to relieve it. The simplicity of the apparatus for opening and shutting the brakes was most beautiful. Opposite the dyna-

From a drawing by R. Dudley.

London. Day & Son, Limited, Lith.

PAYING - OUT MACHINERY

mometer was placed a tiller-wheel, and the man in charge of it never let it go or slackened in his attention for an instant, but watched the rise and fall of the dynamometer as a sailor at the wheel watches his compass. A single movement of this wheel to the right put the brakes on, a turn to the left opened them. A good and experienced brakeman would generally contrive to avoid either extreme of a high or low strain, though there were few duties connected with the laying of submarine cables which were more anxious and more responsible while they last, than those connected with the management of the brakes. The whole machine worked beautifully, and with so little friction that when the brakes were removed, a weight of 200 lb. was sufficient to draw the Cable through it.

In order to guard against any possible sources of accident, every preparation was made in case of the worst, and, in the event of very bad weather, for cutting the Cable adrift and buoying it. For this purpose a wire rope of great strength, and no less than five miles long, having a distinctive mark at every 100 fathoms, was taken in the Great Eastern. This, of course, was only carried in case of desperate eventualities arising, and in the earnest hope that not an inch of it would ever be required. If, as unfortunately happened, its services were wanted, the Cable could be firmly made fast to its extremity, and so many hundred fathoms of the wire rope, according to the depth of water the Cable was in, measured out. To the other end of the rope an immense buoy was attached, and the whole would then be cut adrift and left to itself till better weather.

On the 24th of May, His Royal Highness the Prince of Wales, accompanied by many distinguished personages, paid a long visit to the Great Eastern, for the purpose of inspecting the arrangements made for laying the Cable. His Royal Highness was received by Mr. Pender, the Chairman of the Telegraph Construction Company; Mr. Glass, Managing Director; and a large number of the electricians and officers connected with the undertaking. After partaking of breakfast, the Prince visited each portion of the ship, and witnessed the transmission of a message sent through the coils, which then represented in length 1,395 nautical miles. The signals transmitted were seven words, " I WISH SUCCESS TO THE ATLANTIC CABLE," and were received at the other end of the coils in the course of a few seconds—a rate of speed which spoke hopefully of success.

On Monday, the 29th of May, the last mile of this gigantic Cable was completed at Glass, Elliot, & Co.'s works; an event celebrated in the presence of all the eminent scientific men who had laboured so zealously in the promotion of the undertaking at Greenwich. When the tinkling of the bell gave notice that the machine was empty, and the last coil of the Cable stowed away, the mighty work, the accomplishment of which was their dream by night and their study by day, stood completed. For eight long months the huge machines had been in a constant whirl, manufacturing those twenty-three hundred nautical miles of Cable destined to perform a mission so important, and yet it would be difficult to point to a single hour during which they did not yield something to cause care and anxiety.

On Wednesday, the 14th of June, the Amethyst completed her final visit, and commenced to deliver the last instalment of the Cable to the Great Eastern.

On the 24th the Great Eastern left the Medway for the Nore, carrying 7000 tons of Cable, 2000 tons of iron tanks, and 7000 tons of coal. At the Nore she took in 1,500 additional tons of coal, which brought her total dead-weight to 21,000 tons.

Mr. Gooch, M.P., Chairman of the Great Eastern Company and Director of the Telegraph Construction and Maintenance Company; Mr. Barber (Great Eastern), Mr. Cyrus Field, Captain Hamilton, Directors of the Atlantic Telegraph Company; M. Jules Despescher; Mr. H. O'Neil, A.R.A.; Mr. Brassey, Mr. Fairbairn, Mr. Dudley, the representatives of some of the principal journals, and several visitors, went round in the vessel from the Nore to Ireland.

The whole of the arrangements for paying-out and landing the Cable were in charge of Mr. Canning, principal Engineer to the Telegraph Construction and Maintenance Company, Mr. Clifford being in charge of the machinery. These gentlemen were assisted by Mr. Temple, Mr. London, and eight experienced engineers and mechanists. A corps of Cable layers was furnished by the Telegraph Construction and Maintenance Company.

The Electrical Staff consisted of

C. V. de Sauty .	Chief.
H. Saunders . .	Electrician to the Malta and Alexandria Telegraph.
Willoughby Smith	Electrician to the Gutta Percha Company.
W. W. Biddulph .	Assistant Electrician.
H. Donovan . .	Do.
O. Smith . . .	Do.
J. Clark . . .	Do.
J. T. Smith . .	Instrument Clerk from Malta and Alexandria Telegraph.
J. Gott	Do.　　Do.　　Do.
L. Schaefer . .	Mechanician.

T. Picken lith. from a drawing by R. Dudley.

London, Day & Son, Limited, Lith

COILING THE CABLE IN THE AFTER TANK ON BOARD THE GREAT EASTERN AT SHEERNESS.–VISIT OF H.R.H THE PRINCE OF WALES ON MAY 24TH

The Staff at Valentia was composed of

J. May	Superintendent.
T. Brown . . .	Assistant Electrician.
W. Crocker . . .	Do.
G. Stevenson . .	Instrument Clerk from Malta and Alexandria Telegraph.
E. George . . .	Do. Do. Do.
H. Fisher . . .	Do. Do. Do.

All the arrangements at Valentia were under the direction of Mr. Glass.

Mr. Varley, chief electrician to the Atlantic Telegraph Company, was appointed to report on the laying of the Cable, and to see that the conditions of the contract were complied with. Associated with him was Professor W. Thomson, LL.D., F.R.S., of Glasgow. His staff was composed of Mr. Deacon, Mr. Medley, Mr. Trippe, and Mr. Perry.

Several young gentlemen interested in engineering and science were accommodated with a passage on board.

At noon on July 15th the Great Eastern, in charge of Mr. Moore, Trinity pilot, drawing 34 ft. 4 in. forward, and 28 ft. 6 in. aft, got up her anchor, and at midnight on July 16th was off the Lizard. On Monday, 17th, she came up with the screw steamer Caroline, freighted with 27 miles of the Irish shore end of the Cable, weighing 540 tons, and took her in tow. Then a gale set in, which gave occasion to the Great Eastern to show her fine qualities as a sea-boat when properly handled. Even those who were most prejudiced or most diffident, admitted that on that score no vessel could behave better. This trial gave every one, from Captain Anderson down, additional reason to be satisfied with the fitness of the great ship for the task on which she was engaged. Next day, Tuesday, July 18th, she encountered off the Irish coast a strong gale with high westerly sea, through which she ran at the rate of six knots an hour. The Caroline, which rolled so heavily and pitched so vigorously as to excite serious apprehensions, broke the tow rope in the course of the day, and ran for Valentia harbour, where she arrived safely, piloted by the Great Eastern; and the Great Eastern, passing inside the Skelligs, stood in close to Valentia Lighthouse, and sent a boat ashore to communicate. H.M.S. Terrible, Captain Napier, and H.M.S. Sphinx, Captain V. Hamilton, were visible in the offing, having sailed at the end of the previous week from Queenstown for the rendezvous, outside Valentia. Captain Anderson having fired a gun to announce his arrival, steamed for Berehaven, in Bantry Bay, and anchored inside the island on Wednesday morning, July 19th, in 17 fathoms. Here the Great Eastern lay, preparing for her great errand—perhaps, as it may prove, her exclusive "mission,"—on Thursday, 20th, Friday, 21st, and Saturday, 22nd July, whilst the Caroline was landing the shore end of the Cable in Foilhummerum Bay in Valentia. During her stay in Bantry Bay, many visitors, high and low, came on board the Great Ship, but it was believed all over the country that she was going to Foilhummerum. The greater portion of those anxious to see her made the best of their way to that secluded spot, to which there was once more attached an interest of a civilised character; for, if country legends be true, there must have been some regard paid to Foilhummerum Bay by no less a person than Oliver Cromwell, testified yet by the grey walls of a ruined fort, and traces of a moat and outer wall, on the greensward above the point which forms the northern entrance to the lonely bay. This crisp greensward, glistening with salt, lies in a thin crust over the cliffs, which rise sheerly from the sea some three or four hundred feet; and for what Oliver Cromwell or any one else could have erected a fortalice thereon, may well baffle conjecture, unless the builder, having a far-reaching mind, saw the importance of watching the most westerly portion of Europe, or anticipated the day when Valentia would be recognised as one of the landmarks created by the necessities of commercial and social existence. Taking advantage of the shelter afforded by a gradual descent inland of the soil, a few cabins have been placed by the natives—half-fishermen, half-husbandmen—Archytas-like, spanning land and sea, and making but poor subsistence from their efforts on both. The little bay, which is not much above a mile in length, contracts from a breadth of half so much, into a watery *cul-de-sac*, terminated by steep banks of shale, earth, and high cliff, furrowed by watercourses; and on the southernmost side it is locked in by the projecting ledges of rock forming the northern entrance to the Port Magee channel. It is so guarded from wind and sea, that on one side only is it open to their united action, but as the entrance looks nearly west, the full roll of the Atlantic may break in upon it when the wind is from that point; and indeed there is not wanting evidence that the wild ocean swell must tumble in there with frightful violence. Jagged fragments of masts and spars are wedged into the rocks immovably by the waves, and the cliffs are gnawed out by the restless teeth of the hungry water into deep caves. But then a sea from that point would run parallel with the line of the Cable, and would sweep along with and not athwart its course, so that the strands would not be driven to and fro and ground out against the bottom. Except for a couple of hundred feet near the shore at the top of this cove, indeed, the bottom is sandy, and the rocks inside the sand line were calculated to form a protection to the Cable, once deposited, as the greater part of its course lay through a channel which had been cleared

T. Picken, lith. from a drawing by R. Dudley

London, Day & Son, Limited, Lith

FOLLHUMMERUM BAY, VALENCIA, LOOKING SEAWARDS FROM THE POINT AT WHICH THE CABLE REACHES THE SHORE

of the boulders with the intention of rolling them back again at low water, to cover in the shore end. Lieutenant White, and the hardy and hard-working sailors of the Coastguard Station at Valentia, had been indefatigable in sounding and buoying out a channel from the beach clear out to sea, within which the Caroline was to drop the Cable. A few yards back from the cliff, at the head of the cove, the temporary Telegraph Station reared its proportions in imitation of a dwarf Brompton boiler—a building of wood much beslavered with tar and pitch, of exceeding plainness, and let us hope of corresponding utility. Inside were many of the adjuncts of comfort, not to speak of telegraphic luxury, galvanometers, wires, batteries, magnets, Siemens's and B. A. unit cases, and the like, as well as properties which gave the place a false air of campaigning. A passage led from end to end, with rooms for living and sleeping in to the right and left, and an instrument room at the far extremity. Here, on a narrow platform, were the signal and speaking apparatus connected with the wires from the end of the Cable, which was secured inside the house. Outside the wires were carried by posts in the ordinary way to the station at Valentia, whence they were conveyed to Killarney, and placed in communication with the general Telegraphic system over the world. The Telegraphic staff and operators were lodged in primitive apartments like the sections of a Crimean hut, and did not possess any large personal facility for enjoying social intercourse with the outer world, although so much intelligence passed through their fingers. But Foilhummerum may in time become a place with something more real than a future. If vessels from the westward do not like to make their number outside, there is nothing to prevent their running into Valentia for the purpose, at all events. On the plateau between the station and the cliff, day after day hundreds of the country people assembled, and remained watching with exemplary patience for the Big Ship. They came from the mainland across Port Magee, or flocked in all kinds of boats from points along the coast, dressed in their best, and inclined to make the most of their holiday, and a few yachts came round from Cork and Bantry with less rustic visitors. Tents were soon improvised by the aid of sails, some cloths of canvas, and oars and boathooks, inside which bucolic refreshment could be obtained. Mighty pots of potatoes seethed over peat fires outside, and the reek from within came forth strongly suggestive of whisky and bacon. Flags fluttered—the Irish green, with harp, crown surmounted; Fitzgerald, green with its blazon of knight on horse rampant, and motto of "Malahar aboo"—faint suspicion of Stars and Stripes and Union Jack, and one temperance banner, audaciously mendacious, as it flaunted over John Barleycorn. Nor was music wanting. The fiddler and the piper had found out the island and the festive spot, and seated on a bank, played planxty and jig to a couple or two in the very limited circle formed in the soft earth by plastic feet or ponderous shoemasonry, around which, sitting and standing, was a dense crowd of spell-bound, delighted spectators. In the bay below danced the light canvas-covered canoe or coracle in which the native fishermen will face the mountain billows of the Atlantic when no other boat will venture forth; and large yawls filled with country people passed to and fro, and the bright groupings of colour formed on the cliffs and on the waters by the red, scarlet, and green shawls of the women and girls, lighted up the scene wonderfully.

It would be gratifying if in such a primitive spot one could shut his eyes to the painful evidence that the vices of civilisation—if they be so—had crept in and lapt the souls of the people in dangerous pleasures. But it could not be denied that the spirit of gambling and gourmandise were there. Seated in a ditch, with a board on their knees, four men were playing "Spoil Five" with cards, for discrimination of which a special gift must have been required; but they were as silent, eager, and grave, as though they had been Union or Portland champions contesting last trick and rub. Near them was one who summoned mankind to tempt capricious Fortune by means of an iron skewer, rotating an axis above a piece of tarpaulin stretched on a rude table, which was enlivened by rays of vivid colour. At the end of each ray was an object of art—the guerdon of success—an old penknife, brass tobacco-box, tooth-comb, thimble, wooden nutmeg, or the like. A very scarecrow professor of legerdemain and knavery hid his pea, and challenged detection, and divided public attention with a wizard who presided over a wooden circle with a spinning needle in the centre to point to radii, at end of which were copper moneys deposited by the adventurers, who generally saw them whisked off into the magician's grimy pocket. An ancient woman, spinning, and guarding a basket of most atrabilious confectionery, and a stall garnished with buttons and gingerbread, completed the attractions of Foilhummerum during this festive time.

The matter of wonder was, what the people flocked to see, for it must soon have been known the Great Eastern was not there. The Hawk and the Caroline, as they went into Valentia, did duty successfully for the Big Ship, and the steam-yacht Alexandra, belonging to the Dublin Ballast Board, and H.M. tender Advice, created a sensation as they appeared in the offing on their way to the same rendezvous. All that related to the Cable and the laying of it possessed the utmost interest for the country people, simply because the Cable went westwards across the ocean to the home of their hopes. Many of the poor people believed that it would facilitate communications with their friends in the land to which their thoughts are for ever tending, remembering

T.Picken, lith. from a drawing by R.Dudley.

London, Day & Son, Limited, Lith

THE CLIFFS FOILHUMMERUM BAY, POINT OF THE LANDING OF THE SHORE END OF CABLE. JULY 22ND.

perhaps the words of Lord Carlisle when he told them of the advantages the Telegraphic Cable would confer upon them.

The village of Knightstown witnessed an unusual influx of visitors, and those whom the hospitable roof of Glenleam could not stretch its willing eaves over, found something more than shelter in the inn and in the comfortable houses which acted as its succursales on the occasion. But there was in the midst of all the pleasurable excitement of the moment a tinge of dissatisfaction, because the people had persuaded themselves that if they were not to see the Great Eastern in the harbour, they would at least have H.M.S.S. Terrible and Sphinx, and the satellites of the Leviathan in their anchorage, and all they beheld of the men of war was their smoke and faint outlines on the distant horizon.

The Terrible and Sphinx might have coaled in Valentia, and waited there for the arrival of the Great Eastern, of which they could have heard by telegraph, instead of towing colliers to Cork and going into Berehaven, where there is no telegraph. Now, as to this harbour, let it be admitted at once that its entrance is only 180 yards broad. But the "Narrows" of Valentia Harbour is like a very short neck to a bottle, and after less than a ship's length, the channel enlarges sufficiently to allow several vessels to sail abreast in water which is never rough enough to prevent the passage of boats to Begennis or Renard Point. Indeed, Capt. Wolfe's report to the Hydrographer to the Admiralty expresses an opinion that the Needles' passage is more intricate and dangerous. The Skelligs on one side and the Blasketts on the other mark the approach very distinctly. Inside, there is 600 acres, or more than a square mile, of harbour, with good holding ground, having a maximum of six furlongs and a minimum of three furlongs water.

The disappointment caused by the cautious indifference of the Terrible and Sphinx to the advantages of lying snugly inside Valentia Harbour was felt acutely. The Knight of Kerry, who has taken such an interest in the under-taking, and all the inhabitants, regarded it as a mark of distrust in the safety of the anchorage and in the facility of access to it, which was without any justification, and some ascribed it to less creditable influences and objects; but no one could believe that the officers in command of the ships kept out at sea in such weather, wearying the crews and wasting coals, without direct orders, or that they would hesitate to run in, if left to themselves, as soon as it was evident the point of rendezvous ten miles from shore was not intended as a permanent station. The harbour had been visited by H.M.S.S. Stromboli, Hecate, Leopard, Cyclops, the U.S. frigate Susquehanna, and many large merchantmen, including the Carrier Dove, a vessel of 2,400 tons.

On July 19th a channel was made down the cliff to the beach for the shore end of the Cable, which was carried down in an outer case through a culvert of masonry, and deposited in a cut made as far into the sea as the state of the tide would admit. On the 21st an "earth" Cable, with a zinc earth, on Mr. Varley's plan, was carried out into the bay from the station, and safely deposited outside the channel marked for the Cable. The Caroline went round from Valentia to Foilhummerum, and on July 22nd the shore end of the Cable was carried from her over a bridge formed of twenty-five yawls belonging to the district, amid great cheering, and hauled up the cliffs to the station. The safe arrival of the terminal wire in the building, in the presence of a large assemblage, took place at 12·45, Greenwich time, and as the day was fine, the scene, to which the fleet of boats in the bay gave unusual animation, was witnessed to the greatest advantage.

When the excitement caused by the landing of the Cable was abated, the Knight of Kerry was called on to speak to the people assembled outside the Instrument Room, and said :—" I feel that in the presence of so many who have taken an active and a useful part in this undertaking, it may seem almost presumptuous in me to open my mouth on this occasion ; but from the very beginning I have felt an interest which I am sure the humblest person here has also felt in the success of this the greatest undertaking of modern times. I believe there never has been an undertaking in which, not to speak disparagingly of the commercial spirit and the great resources and strength of the land, that valuable spirit has been mixed up with so much that is of a higher nature, combining all the most noble sentiments of our minds, and the feelings intended for the most beneficial purpose, which are calculated to cement one great universe, I may say, with another. I do not think we should be quite silent when such an undertaking has been inaugurated. It has been discussed whether this ceremony should be opened with a prayer or not. Whether that shall be done or not, I am sure there is not a person present who does not feel the utmost thankfulness to the Giver of all Good for having enabled those who have taken an active part in it to bring this great undertaking to what I am sure will have a happy issue. I do not think anything could be fitly added to the sentiment of the first message which was conveyed, namely — ' Glory to God in the highest, on earth peace, good will toward men.' I shall not detain you with another word, but will only ask you all to give the heartiest cheers for the success of the undertaking. I will also take the liberty of asking you, when you have done that, to give three cheers for a gentleman who has come here at great inconvenience, and has done us very great honour in doing so, and who deserves them, not only from his position and character, but also from

G. McCulloch lith. from a drawing by R.Dudley

London. Day & Son, Limited, Lith

FOILH MMFRUM BAY, VALENCIA FROM "CROMWELL FORT" THE CAROLINE AND BOATS LAYING THE EARTH WIRE JULY 21ST

the interest which he has always shown in this undertaking. I call upon you to give three hearty cheers for Sir Robert Peel."

The meeting responded very heartily to the call, and when silence was restored, Sir Robert Peel said : " Gentlemen, as the Knight of Kerry has well observed, this is one of the most important works that this country could have been engaged in, inasmuch as it tends to draw us together in a link of amity and friendship with a mighty continent on the other side of the Atlantic. I trust, as the Knight of Kerry has so justly observed, that it may tend not only to promote the peace and commerce of the world, but that it may also lead to a union of feeling and to good fellowship between those two great countries ; and I trust that as it has been so happily inaugurated to-day, so it may be successful under the exertions of those who have taken part in it to-day and for some time past. Gentlemen, I think the progress of this undertaking deserves that we should pay the highest compliment to those who have been actively engaged in carrying it out to the stage at which it has arrived. We are about to lay down, at the very bottom of the mighty Atlantic, which beats against your shores with everlasting pulsations, this silver-toned zone, to join the United Kingdom and America. Along that silver-toned zone, I trust, may pass words which will tend to promote the commerce and the interest of the two countries ; and I am sure we will offer up prayers for the success of an undertaking, to the accomplishment of which persevering industry and all the mechanical skill of the age have been brought to bear. Nothing has been wanting in human skill, and therefore for the future, as now, let us trust the hand of Divine Providence will be upon it ; and that as the great vessel is about to steam across the Atlantic no mishaps or misfortune may occur to imperil or obstruct the success of the work which has now been so happily commenced. I ask you all to give a cheer in honour of my noble friend here, the Knight of Kerry, who has just begun the work."

The demand was enthusiastically complied with, for the Knight is an immense favourite with all the dwellers in his little dominion.

Sir Robert Peel then said : " Now, gentlemen, probably one of the first messages that will be sent by this Cable will be a communication from the Sovereign of this great country to the great ruler of the mighty continent at the other side of the Atlantic. I will ask you to give three cheers for her Majesty the Queen." (Cheers.) Sir Robert Peel in conclusion, said : " I give you, with hearty good will, health and happiness to the ruler of the United States, President Johnson." (The toast was received with loud cheers.)

Mr. Glass, who was called on to acknowledge the hearty reception given to his name and the Company's, said : " On behalf of myself and those connected with

H

me in this undertaking, I beg to return you thanks. I am glad that our labours have been appreciated by those around us. I assure you that the work that has been so far completed has been a source of great anxiety to us all; but that anxiety has been relieved very much by the fact that we have now landed a Cable which we one and all believe to be perfect. I believe that nothing can interfere with the successful laying of the Cable but the hand of the Almighty, who rules the winds and waves. So far as human skill has gone, I believe we have produced all that can be desired. We now offer up our prayers to the Almighty that He will grant success to our undertaking."

The Doxology was then sung, with which this part of the proceedings closed, and the electricians busied themselves with securing the shore end confided to their charge in its new home.

At 2 o'clock in the afternoon the Caroline, towed by the Hawk, and attended by the Princess Alexandra and Advice, proceeded to sea, veering out the shore end of the Cable in the channel marked by Lieutenant White, and at 10·30 p.m. buoyed the end 26 miles W.N.W. of Valentia, in 75 fathoms of water. A message was sent through the Cable to Foilhummerum, and a dispatch was forwarded to the Great Eastern, in Bantry Bay, to come round with all speed. This order was obeyed with such diligence that her appearance off the harbour of Valentia was reported in Knightstown soon after 7 o'clock next morning, July 23. H.M.S. Terrible and H.M.S. Sphinx were in company. The Hawk, which returned from the Caroline in the course of the night, got up steam and left Valentia Harbour about 10 o'clock a.m., July 23, with a party of visitors and passengers for the Great Eastern, among the former being Sir R. Peel, the Knight of Kerry, and Captain Lord John Hay. By 3 p.m. the Hawk had reached the flotilla, which lay around the buoy, preparing for the great enterprise. She was just in time; the end of the shore Cable was about to be spliced and joined with the landward end of the main Cable from the after tank of the Great Eastern, and the boats of the Great Ship and of the two men-of-war, were engaged in carrying the end of the main Cable to the Caroline. Sir R. Peel, the Knight of Kerry, Lord John Hay, Mr. Canning, and others, got on board the Great Eastern in successive trips of the Hawk's boats; but the ladies, who had come so far and had suffered too in order to see the famous vessel, could not venture, as there was a swell on which made it difficult to embark or approach the gangway ladders. After an hour's enjoyment of the almost terrestrial steadiness of the Great Eastern, the visitors departed, amid loud cheers, to the Hawk, and at 5·10 p.m. it was reported by the electricians that the tests of the splice between the main Cable and the shore end were complete, and that the shore end was much improved in its electrical condition by its immersion in the water. The boats were hoisted in by the men-of-war and by the Great Eastern, adieux and good wishes were exchanged, and, with hearts full of confidence, all on board set about the work before them.

The bight of the Cable was slipped from the Caroline, at 7·15 p.m., and the Great Eastern stood slowly on her course N.W.¼W. Then the Terrible and Sphinx, which had ranged up alongside, and sent their crews into the shrouds and up to the tops to give her a parting cheer, delivered their friendly broadsides with vigour, and received a similar greeting. Their colours were hauled down, and as the sun set a broad stream of golden light was thrown across the smooth billows towards their bows as if to indicate and illumine the path marked out by the hand of Heaven. The brake was eased, and as the Great Eastern moved ahead the machinery of the paying-out apparatus began to work, drums rolled, wheels whirled, and out spun the black line of the Cable, and dropped in a graceful curve into the sea over the stern wheel. The Cable came up with ease from the after tank, and was payed-out with the utmost regularity from the apparatus. The system of signals to and from the ship was at once in play between the electricians on board and those at Foilhummerum. On board there were two representative bodies —the electricians of the Telegraph Construction and Maintenance Company, under M. de Sauty, and the electricians of the Atlantic Telegraph Company, Mr. Varley, Professor Thomson, and assistants. The former were to test the electrical state of the Cable as it was being payed-out, and to keep up signals between the ship and the shore. The latter, who had no power of interference or control, were simply to report on the testing, and to certify, on their arrival in New-foundland, whether the Cable fulfilled the conditions specified in the contract. The mechanical arrangements for paying-out the cable were in charge of Mr. Canning, engineer-in-chief to the Telegraph Construction and Maintenance Company, who might be considered as having supreme control over the ship *ad hoc*. In the space on deck between the captain's state-room and the entrance to the grand saloon, was the Testing-Room—a darkened chamber, into which were led conducting wires from the ends of the Cable, for the ordeal to which they were subjected by the electricians, at a table whereon were placed galvanometers and insulation and resistance-testing machines.

The instructions for signalling, determined upon by the electricians of the Telegraphic Construction and Maintenance Company, were as follows:—

1. During the paying-out of the Cable, from the moment of starting until the end is landed at Newfoundland, electrical tests will be applied without intermission.

2. The tests will be for insulation, for continuity, and to determine the resistance of the conductor, the whole length of Cable being joined up in one length.

3. Each series of tests will commence at the hour (Greenwich time), and will last one hour.

4. The insulation test will consist of 30 minutes' electrification of the Cable, commencing at the hour, and lasting till 30 minutes past the hour. Readings of the galvanometer to be taken every minute, commencing one minute after contact with the battery, the battery to consist of 40 cells.

5. At 30 minutes past the hour signals will be received from the shore for 10 minutes. Unless the ship wishes to communicate with shore by special speaking instruments, in which case, instead of receiving signals from the shore, ship will put on a C to E current to oppose deflection on shore. Galvanometer to arrest shore attention, and when joined, give the call as in paragraph 9 : the ordinary signals will be 5 reversals of 2 minutes each.

6. At 40 minutes, C of Cable will be taken to 10 minutes.

7. At 50 minutes signals will be sent to the shore, and for the ordinary signals 5 reversals, 2 minutes each, commencing C to E.

8. Then a repetition of the same tests to be made and continued without any interval.

9. In case it becomes necessary to speak to shore by speaking instruments, the signal will be given at the 50 minutes, and at the 30 minutes, as in paragraph 5, by sending $8\frac{1}{4}$ minutes' reversals, commencing Z to E, and changing over to the speaking instruments, on receiving acknowledgment of call from shore (which will be also $8\frac{1}{4}$ minutes' reversals), communication or message to be sent, and when acknowledgment of message and reply (if any) is received, then the system of testing is to be resumed, as if no interruption had taken place.

10. Every 50 nauts. of Cable payed-out will be signalled at the same time (viz., at the 50 mins.), thus, instead of 5 reversals of 2 minutes, 10 reversals of 1 minute will be made commencing Z to E.

11. Every 50 nauts. distance run will be signalled to the shore ; the signal will be 2 reversals (commencing Z to E), each 2 minutes' duration—2 reversals, each 1 minute's duration, and 2 reversals, each 2 minutes' duration.

12. Should any defect in signals be perceived, or bad time kept, notice will be given to the shore by signalling at the 50 minutes—thus, by giving 2 reversals of 5 minutes' duration, commencing Z to E.

13. In sounding, signal will be one current of 10 minutes' duration, Z to E.

14. Land-in-sight signal will be likewise one current of 10 minutes' duration, Z to E.

15. Greenwich time will be kept, but a column will be devoted in journals and sheets to ship's time.

16. After the insulation test is taken, it is to be worked out thus—The same deflection at the 15th minute's reading will be obtained with the same battery through resistance, and a shunt to the galvanometer. The amount of resistance multiplied by multiplying power of the shunt, and galvanometer multiplied by the length of the Cable, will give the G. p. R. pr. nt.

T. Picken, lith, from a drawing by R. Dudley.

London. Day & Son, Limited, Lith

THE GREAT EASTERN UNDER WEIGH, JULY 23RD (ESCORT AND OTHER SHIPS INTRODUCED BEING THE TERRIBLE, THE SPHINX, THE HAWK & THE CAROLINE)

17. The copper resistance of the Cable will be taken after 5 minutes' electrification.

18. No change in the instruments, wires, or connections (other than the batteries, if necessary), to be made on any account, unless such instruments, &c., become defective—any necessary change to be made as quickly as possible.

19. Should the rolling of the ship generate a magnetic current of sufficient strength to embarrass the signals, a stronger current for the signals will be put on on shore, and a shunt used with the galvanometer on board, notice to the shore to put on more power will be given by one current of 5 minutes, commencing Z to E, and 5 reversals of 1 minute's duration.

20. The iron earth of the Cable will be used both on board and on shore—other earths, however, to be in readiness for use, if necessary.

21. Full particulars of every test and every occurrence in the testing-room to be entered in journal, together with the name of the electricians on duty, and the time of their coming on and going off duty.

22. After the end is landed, should signals fail, the paying-out system to be resumed until signals are re-established.

23. In case of a minute fault appearing, such as will partially affect the signalling, but which will not stop the communication entirely, notice will be given to shore to reduce battery power. Such notice will be given at the 50 minutes, by sending 5 reversals of 1 minute each, commencing Z to E, and 1 current of 5 minutes' duration.

24. A proper supply of lamps, glasses, oil, and wicks; instrument ink and instrument paper, in sufficient quantities; paraffin, wicks, and spare lamp-glasses for the instrument lamps; lamp-brushes, tools, sulphate of copper, stationery, &c., to be always ready for use.

25. No person except those on duty, and the engineers and the officers authorised by the Atlantic Telegraph Company, to be allowed in the instrument room on any pretence.

26. The batteries to be kept in an efficient state, especially those for sending reversals —their force taken periodically, and if any variety occur, they must be renewed, or brought up to the original force.

27. Supplies of every material needful for such purpose to be in constant readiness.

28. The actual end of the Cable to be brought to the instrument tables, and well insulated.

SHIP'S SIGNALS.

29. Ordinary.—5 reversals, commencing C to E, each 2 minutes.

To open communication.—8 reversals, commencing Z to E, each ¼ minute.

50 nauts. payed out.—10 reversals, commencing Z to E, each 1 minute.

50 nauts. distance run, signal will be, 2 reversals, each 2 minutes, commencing Z to E.

   „   „   „   2   „   „   1   „   „   „

   „   „   „   2   „   „   2   „   „   „

Defective signals.—2 reversals, commencing Z to E, each 5 minutes.

In soundings.—1 current of 10 minutes, Z to E.

Land in sight.—1   „   „   „   „

Notice to increase power.—1 current of 5 minutes, commencing Z to E, and 5 reversals of 1 minute's duration.

Notice to reduce power.—5 reversals of 1 minute, commencing Z to E, and 1 current of 5 minutes.

SHORE.

1. During the paying-out of the Cable, from the moment of starting until the end is landed at Newfoundland, a system of testing will be applied without intermission.

2. The tests will be for insulation, for continuity, and to determine the copper resistance of the conductor.

3. Each series of tests will commence at the hour (Greenwich time), and will last 1 hour. Both the insulation and C R tests will be made on board.

4. The insulation test will be made on board, and to enable that to be done, the end of the Cable must be insulated on shore for 30 minutes, commencing at the hour.

5. At the 30 minutes past the hour, signals will be sent to the ship for 10 minutes. Should ship at this time desire to open communication, ship will put on a current so as to oppose shore's current on his galvanometer, to arrest shore's attention, and will, when gained, give the call as in paragraph 10.

6. The ordinary signal will be 5 reversals of 2 minutes' duration, commencing C to E.

7. At the 40 minutes, Cable to be put to earth direct, without any instrument being in circuit.

8. At the 50 minutes, signals will be received from the ship. The ordinary signal will be 5 reversals, each 2 minutes' duration.

9. Then a repetition of the same series to be made and continued.

10. Should ship desire to open communication by special speaking instruments, notice will be received by a signal of 8 reversals (giving a deflection the opposite to the ordinary signals) of $\frac{1}{4}$ minute's duration.

11. After returning the same signal to the ship as an acknowledgment, the speaking instruments to be put in circuit, and the message from the ship received, and when acknowledgment of message, or reply, is given, the regular system of signals to be resumed as if no interruption had occurred.

12. Every 50 nauts. of the Cable payed-out will be signalled to the shore by signal (instead of the ordinary signals). This signal will be 10 reversals of 1 minute each— the first current giving a deflection the opposite side to the first current of the ordinary signals.

13. Every 50 nauts. distance run will be signalled to the shore : the signal will be 2 reversals of 2 minutes' duration, 2 reversals of 1 minute's duration, and 2 reversals of 2 minutes' duration—the first current giving a deflection opposite to the first deflection of the first current of the ordinary signal.

14. Should ship receive weak or defective signals, or bad time kept, notice will be given by sending 2 reversals of 5 minutes each, commencing the opposite side to the ordinary signals.

15. When the ship gets into soundings, notice will be given by sending one current of 10 minutes' duration, the opposite side to the first current of the ordinary signals.

16. When land is in sight, notice will be given by the same signal.

17. Greenwich time to be kept, but a column to be devoted to local time in the journals and sheets.

18. No change in instruments, wires, or connections (other than the batteries, if necessary), to be made on any account, unless such instruments become defective, and any necessary change to be made as quickly as possible.

19. Should the rolling of the ship generate a magnetic current of sufficient strength to embarrass the signals, a stronger current for the signals must be put on by shore on receiving notice from the ship; the notice will be given by 1 current of 5 minutes', and 5 reversals of 1 minute's duration.

20. The iron earth of the Cable to be used both on board and on shore : copper earths, however, will be in readiness for use if necessary.

21. Full particulars of every occurrence in the testing-room will be entered in journals, together with the names of the electricians on duty, and the time of their coming on and going off duty.

22. When the end is landed at Newfoundland, should signals fail at any time, the paying-out system to be resumed until signals pass again freely.

23. On receiving a signal of 5 reversals of 1 minute's, and a current of 5 minutes' duration, shore must reduce the battery power used for sending reversals by one-half, and on a repetition of the same signal again reduce the power one-half, until (should notice continue to be given to that effect) the minimum of power be reached.

24. Shore must not have the privilege of opening a conversation, or to use or call for the use of the special speaking instruments, under any circumstances, except to give notice of any accident that may cause an interruption of signals, or that may affect the safety of the Cable or signals.

25. Should any interruption of signals from the ship occur by reason of an accident on board, shore will continue to free the Cable at the usual time, and to put to earth direct at the usual time, and in the intervals to put into circuit with the Cable a galvanometer, and watch the same for signals, and continue doing so until communication with the ship is restored, or information is received by other means from the ship.

26. On re-establishment of communication, shore must not ask any questions, but take the resumption of signals as an indication of all being well again, and will continue to follow the series of tests as if nothing had happened.

27. Shore will take time from the ship; should any irregularity in the reception of signals from the ship occur, such irregularity must be entered in journals, and must not form a ground for shore's altering his time, but shore must follow blindly every change (should one take place), as if the most correct time had been kept.

28. A proper supply of lamps, glasses, oil, and wicks ; instrument ink and instrument paper, in sufficient quantities ; paraffin, wicks, and spare lamp-glasses for the instrument lamps ; lamp-brushes, tools, sulphate of copper, stationery, &c., to be always ready for use.

29. No person, except those on duty, and the officers authorised by the Atlantic Telegraph Company, to be allowed in the instrument room on any pretence.

30. The batteries to be kept in an efficient state, especially those for sending reversals —their force taken periodically, and if any variation occur, they must be renewed, or brought up to the original force.

31. Supplies of all materials necessary for such purpose to be in constant readiness.

32. The actual end of the Cable to be brought to the instrument tables, and well insulated.

SHORE SIGNALS.

33. Ordinary.—5 reversals, each two minutes, commencing C to E.

34. To open communication on acknowledgment.—8 reversals, each $\frac{1}{4}$ minute, commencing Z to E.

As the voyage of the Great Eastern promised to be so interesting to electricians and engineers, several young gentlemen who worked in the testing-room and in the engineer's department received a passage, as we have mentioned, but there was no person on board who was not in some way or other engaged on the business of both companies, or connected with the management of the ship. The voyage commenced most favourably. The rate of speed was increased to 3 knots, then to 4 knots, then to 5 knots, and finally, to $6\frac{1}{2}$ knots an hour, and the Cable flew from each coiled flake as if it were eager to push up through the controlling bands of the so-called crinoline, and to plunge into the sea. At 10·49 p.m., Greenwich time, 50 miles of Cable had been payed-out, and the process continued to midnight with equal ease and regularity. In order to make each day's proceedings distinct, and to take the reader over the course so that he can follow the expedition readily by the aid of the accompanying chart, I propose recording events in the form of a diary.

Monday, July 24th.—The morning was exceedingly fine, and the ship proceeded steadily at an average rate of 6 knots an hour, with a light favouring wind and a calm sea. Those who were up betimes had just taken a turn or two on deck, watching for the early dawn, when they observed some commotion in the neighbourhood of the Testing-Room, and soon afterwards the ship's engines were slowed and stopped. According to Professor Thomson's galvanometer, which is used in the system employed in testing, a ray of light reflected from a tiny mirror suspended to a magnet travels along a scale, and indicates the resistance to the passage of the current along the Cable by the deflection of the magnet, which is marked by the course of this speck of light. If the light of the mirror travels beyond the index, or out of bounds, an escape of the current is taking place in the Cable, and

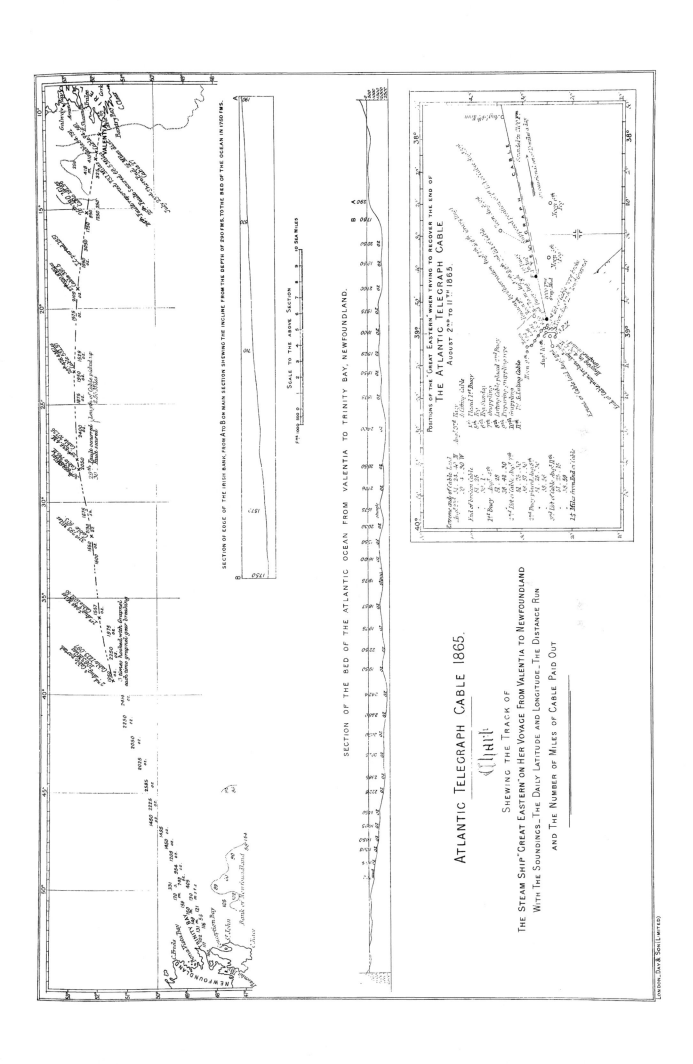

ATLANTIC TELEGRAPH CABLE 1865.

CHART

SHEWING THE TRACK OF

THE STEAM SHIP "GREAT EASTERN" ON HER VOYAGE FROM VALENTIA TO NEWFOUNDLAND

WITH THE SOUNDINGS — THE DAILY LATITUDE AND LONGITUDE — THE DISTANCE RUN

AND THE NUMBER OF MILES OF CABLE PAID OUT

SECTION OF THE BED OF THE ATLANTIC OCEAN FROM VALENTIA TO TRINITY BAY, NEWFOUNDLAND.

SECTION OF EDGE OF THE IRISH BANK, FROM A TO B ON MAIN SECTION SHEWING THE INCLINE FROM THE DEPTH OF 290 FMS. TO THE BED OF THE OCEAN IN 1750 FMS.

SCALE TO THE ABOVE SECTION

POSITIONS OF THE "GREAT EASTERN" WHEN TRYING TO RECOVER THE END OF

THE ATLANTIC TELEGRAPH CABLE

AUGUST 2ND TO 11TH 1865.

LONDON, DAY & SON (LIMITED)

what is technically called a fault has occurred. At 3·15 a.m., when 84 miles of Cable had been paid out, the electrician on duty saw the light suddenly glide to the end of the scale, and then vanish. The whole staff were at once aroused— the news soon flew through the ship. After testing the Cable for some time by signalling to and from the shore, Mr. de Sauty satisfied himself that the fault which had occurred was of a serious character, and measures were taken accordingly to rig up the picking-up apparatus at the bow, to take in the Cable till the defective portion was reached and cut out. Such an early interruption to our progress caused a little chagrin, but the veterans of submarine telegraphy thought nothing of it. Whilst the electricians were testing, to obtain data respecting the locality of the fault, the fires were got up in the boilers of two small engines on deck to work the picking-up machinery. At 4 a.m. a gun was fired by the Great Eastern to call the attention of the Terrible and Sphinx to our proceedings, and they were also informed by signal of the injury. Notwithstanding the skill and experience of the scientific gentlemen on board, there was a great vagueness of opinion among them respecting the place where the fault lay. Some believed the defective part was near the shore, and probably at the splice of the shore end with the main Cable ; others thought it was eastward or westward of the same place ; and calculations, varied by uncertain indications given by the currents showing that the fault itself was of a variable character, and permitted the currents of electricity to escape irregularly, were made by the scientific staff, which fixed it at points from 22 to 42 miles—one at 60 miles—from the ship. But repeated observations gave closer results. Mr. Varley came to the conclusion that the fault was not very far from the ship; and Mr. Sanders, a gentleman who had much experience in fault-finding, arrived at the conviction that it was not more than 9 or 10 miles astern.

The best test taken by Mr. Saunders, 1·30 a.m., Greenwich time, July 25, after the Cable had been cut down to 78·5 miles, gave—

Resistance, shore end disconnected, 2,600 units.
    ,,   ,,   to earth,   312 ,,

Let a and b be the lengths of Cable-conductor, having resistances equal to the first and second of these numbers; l the length of Cable, and D the distance of the fault. The ordinary formula gives

$$D = b - \sqrt{(a-b)(l-b)}$$

Hence, l being 78·5, and a and b being calculated from the observed copper-resistance of the conductor in the after-tank, and various assumed tempeatures of the sea, we should have, were the measurements perfect, results as follows :—

Copper resistance of Cable in after tank, per nautical mile, observed 4·44 units at 61° temperature.	Distances of the fault calculated accordingly from end in ship, when cut at 78·5 miles of cable from shore end.
Hence 4·42 units at 59° temperature	. . . . 6·7 miles.
4·37 ,, 53° ,,	. . . . . 10·1 ,,
4·25 ,, 40° ,,	. . . . . 22·0 ,,
4·02 ,, 35° ,,	. . . . . 27·2 ,,

This would give 22 miles for the most probable distance of the fault, as 40° is the most probable mean temperature of the first submerged length of 75 miles. The true distance proved to be very nearly 3 miles. The discrepance is owing partly of course to want of absolute accuracy in the measurements, but probably more to the variation of the resistance of the fault during the interval between the two measurements.

Iron chains were lashed firmly to the Cable at the stern, and secured to the wire rope carried round outside the ship to the picking-up apparatus at the bows. As the paying-out stopped, a strain came on the Cable, which was down in 400 fathoms of water, and it needed nice management to keep the ship steady, as she had no steerage way. The Cable, having been shackled and secured, was severed at 8·50 a.m., and flew with its shackling into the sea, plump astern. The stoppers which held the wire rope were released, and the rope was payed-out rapidly as the Cable sunk, in order that the ship's head might be brought round, if possible, so as to take the Cable in over the bows in a straight line with its course.

The Great Eastern dropped to leeward when her engines stopped. When the end of the Cable was got in over the bows, and the picking-up engine was set to work, it was discovered that the locomotive boiler intended to keep up a head of steam for the machinery, was defective. Steam was then supplied by one of the boilers of the ship: the drums and wheels of the picking-up machinery began to revolve, slowly dragging in the Cable over the bows, with a strain which at times rose from 10 cwt. to 30 cwt., leaving a very large margin before the breaking point was reached. The ship's bows were kept up to the line of the Cable with great cleverness, and Mr. Canning and his assistants were perfectly satisfied with their progress. It would be too much to expect that all on board should be so easily contented; for in fact the process of picking-up is of the slowest—a mile an hour was considered to be a fair rate of speed, and a mile and a-quarter was something to be very thankful for. Still, the prospect of returning to Ireland and getting back to the shore end, at the highest of these retrogressive celerities, did not prove attractive. Our position, by observation at noon, was Lat. 52° 2′ 30″, Long. 12° 17′ 30″. As the Cable was in fair working order, Mr. Canning transmitted a message to Mr. Glass at Knights-

T. Picken, lith, from a drawing by R Dudley.

London, Day & Son, Limited, Lith

SPLICING THE CABLE (AFTER THE FIRST ACCIDENT) ON BOARD THE GREAT EASTERN_JULY 25TH

town, to send out the Hawk, in order that he might return in her, and ascertain if the shore end of the Cable were defective. If that were not the case, he proposed to sacrifice the portion of Cable already laid, to return and make a new splice of the main line with the shore end, and to start afresh. In the course of the evening a message was received from Mr. Glass, informing Mr. Canning that the Hawk should be sent out as soon as she had coaled the Caroline. The Terrible sent her First Lieutenant, Mr. Prowse, on board, to see if she could render us any assistance. The Sphinx was busied in taking soundings all round the ship, which showed depths varying from 400 to 480 fathoms. The operation of picking up proceeded all day and all night—the weather being fine but cloudy.

Tuesday, July 25th.—The Hawk was observed soon after daybreak coming towards the Great Eastern. The wind was still light and the sea moderate. All during the night the process of picking up was carefully carried on, the Big Ship behaving beautifully, and hanging lightly over the Cable, as if fearful of breaking the slender cord which swayed up and down in the ocean. Indeed, so delicately did she answer her helm and coil in the film of thread-like Cable over her bows, that she put one in mind of an elephant taking up a straw in its proboscis. At 7·15 a.m., Greenwich time, $9\frac{1}{2}$ miles of Cable had been picked up from the sea, and the thin greyish coating of mud which dropped from it showed that the bed of the Atlantic here was of a soft ooze. The Cable had been cut twice on board, to enable the electricians to apply tests separately to the coils in the tanks. At 9 a.m., ship's time, when somewhat more than $10\frac{1}{4}$ miles had been hauled in, to the joy of all the "fault" was discovered. The Cable came in with flagrant evidence of the mischief. The cause of so much anxiety, delay, and bitter disappointment turned out to be a piece of wire of the same kind as that used in the protecting strands of the Cable itself. It was two inches long or so—rather bent in the middle, with one end sharp and bright, as if from a sharp fracture or being cut by a pair of pliers—the other end blunt and jagged. This piece of wire had been forced through the outer covering of the Cable into the gutta percha, so as to injure the insulation, but no one could tell how it got into the tank. The general impression was, that it was a piece of Cable or other wire which had been accidently carried into the tank, and forced into the coil by the pressure of the paying-out machinery as the Cable flew between the jockey-wheels.

Measures were at once taken to make a new splice and joint, rejecting the Cable picked up, a good deal of which had been strained in the process. Signals were made to the fleet that the enemy had been detected, at 9 a.m., and the Terrible replied, " I congratulate you." First a splice was made in the Cable where it had been cut, for the purpose of testing between the after and fore tanks, and all admired the neatness and strength with which it was performed—the conducting wires soldered and lapped over—the gutta percha heated and moulded on the junction; and, finally, the strands carried over the core and secured. During the operation the Hawk returned to Valentia with our letters, and with the good news, which, however, must have been anticipated by the Cable itself. The splice and joint of the end of Cable towards the shore and the end from the after tank was next made. Then these splices were carefully tested and found perfect, and the stream of electricity was once more sent direct to Valentia. After a detention of some twelve hours, the paying-out machinery was again put in action, and the Cable glided out rapidly astern. All seemed to go well. About half a mile of wire had been paid out, when suddenly all communication between the shore and ship ceased altogether! From great contentment there was sudden blank despair! The operators were in consternation. The news spread from end to end of the ship, which again lay in restless quiet on the waters. The faces of the most cheerful became overcast—gloomy forebodings filled men's minds all at once. Why had the Hawk been sent back? Why were not more tests made before she left? Away worked the electricians in their room, connecting and disconnecting, putting in and taking out stops—intensifying and reducing currents. Not a sign! Not a shadow of a sign! Mr. de Sauty suggested they had got hold of the wrong wires, and professors opined that the operators had done wrong in spending time over the splice between the two tanks at the critical moment when they should have been watching the signals from the shore. Anxious groups gathered round the Testing-Room, and the bolder popped in their heads, as if they could learn anything from the dumb mute wires and the clicking of the chronometers, or from the silent operators who bent over the instruments. At 3·15 p.m. the Cable between the two tanks was again cut, and examination was made to make sure no error had been made in the communications. Again the wearisome energy of the picking-up apparatus was to be called into play—once more the Cable was to be shackled and thrown overboard, and hauled up to the bows and pulled out of the water. Such a Penelope's web in 24 hours, all out of this single thread, was surely disheartening. The Cable in the fore and the main tanks answered to the tests most perfectly. But that Cable which went seaward was sullen, and broke not its sulky silence. Even the gentle equanimity and confidence of Mr. Field were shaken in that supreme hour, and in his heart he may for a moment have sheltered, though he did not nurture, the thought that the dream of his life was indeed but a chimæra. Who could bear up against a life of picking-up? And our paying-out seemed to have such an undue share of the reverse process attached to it! But there was a change in the fortunes of the ship and of its freight. The index light suddenly reappeared on its path in the Testing-Room, and the wearied watchers were gladdened by the lighting of the beacon of hope once more. Again there was one of those mutations to which the flesh of submarine telegraph layers is born heir, and after a few moments of breathless solicitude, it was announced that the signals between the ship and the shore had been restored, and that every instant developed their strength. Mr. de Sauty came out of the Testing-Room to inform Professor Thomson of the fact, and Mr. Canning's operations at the bows of the ship for picking up were most gratefully suspended by the intelligence that the machinery would not be required. At 4·15 p.m. the ship steamed on ahead again, and the Terrible and Sphinx were signalled to come on, 37 hours and 10 minutes having been lost by the fault, and consequent detentions. Our position, at noon was found to be, Lat. 51° 58′, Long. 12° 11′ ; total distance from Valentia, 66½ miles ; total Cable payed-out 74 miles (per centage of slack being 14 miles), distance from Heart's Content, 1,596 miles. The communication with shore continued to improve, and was, in the language of telegraphers, O. K. The alternations of hope and fear to which we had been exposed were now pleasantly terminated for the evening, and the saloon became the scene of joyous and animated conversation, and of a good deal of scientific discussion, till the approach of midnight.

The cause of the detention was argued fully, but it was not easy to determine how it came to pass the signalling had been interrupted ; it was generally accounted for by the supposition that the order of the tests had become deranged whilst the splices were being made on board, and some of the electricians were inclined to think that the system was defective, because the intervals were so long that the fault might be overboard some time before it could be detected.

As the sea and wind rose a little, the speed of the ship was diminished from 6½ knots to 5 knots, at which rate the Cable ran out beautifully throughout the night.

July 26th.—The course of the Cable ran smoothly all throughout the night. At 8 a.m. the Great Eastern was 150 miles from Valentia, and 161½ miles of Cable, including the shore end, had been laid—the loss by slack being only 7·63 per cent. The morning was hazy, and a strong wind from the north-west brought up rather a heavy sea, but the Great Eastern was as steady as a Thames steamer : indeed the stability of the vessel was a never-ending theme of admiration. Our consorts were not so indifferent to the roll of the Atlantic. The Terrible thumped through the heavy sea, and buried her bows in foam with dogged determination.

The Sphinx gave very unmistakable indications of having a harder enigma than she bargained for, as she engaged in her task of taking soundings, which now had become important. We were getting into deep water, having passed the bank on which there is only 200 fathoms, and had come suddenly to the slope beginning with 700 fathoms, and running in one degree to 1,750 fathoms. This slope is not, however, severer than that of Holborn-hill, though it looks very severe upon the map. Towards noon the sea and wind increased. The Sphinx, which first sent down topgallant masts, finally sent down topmasts, but was unable to make head in the sea way, and dropped further and further astern. At noon our course was W.N.W. $\frac{3}{4}$ W., the wind being strong on the port bow, and the weather thick all round, with drizzling mist. Our position was made out to be Lat. 52° 18' 42", Long. 15° 10", distance run 111$\frac{1}{2}$ miles, Cable paid out 125 miles, total distance from Valentia 178 miles. At 1·45 p.m. the Terrible signalled that the Sphinx was unable to keep up with us, but the Cable was running so easily it was resolved not to diminish our speed. Later in the afternoon, the Terrible sent down topgallant masts; later again, she signalled that we were going too fast for the Sphinx; but as the Great Eastern was not exceeding 6$\frac{1}{2}$ knots an hour, at which rate the Cable rolled off easily from the drums, the engineers did not think it advisable to reduce her speed, and so the Sphinx was left further astern, till at length she was hull down on the grey horizon. Each hour it became more important to know what depth of water we were in; and the inconvenience of parting with the Sphinx was felt, as well, perhaps, as the defective nature of the arrangements with the Admiralty, which had furnished only one sounding apparatus. The Terrible had got no deep-sea sounding apparatus. There was none on board of the Great Eastern. In deep-sea soundings a special apparatus is requisite, and the leads and the lines ordinarily used by men-of-war only penetrate the upper strata of the waters of the Atlantic. It was conjectured that we had passed over the 2,050 fathoms' soundings, and the Cable proved, by a slightly increased pressure on the dynamometer, that its trail was lengthening in the watery waste ere it ruffled the smooth surface of the ooze two miles below. The insulation tests showed an improvement, and the transmission of signals between the ship and the shore afforded most satisfactory indications. At night the wind came round to the north-west, the sea somewhat decreased, and as evening closed in, the Terrible drew up on our beam, working two boilers; but when night fell, the Sphinx was scarcely visible on the distant horizon.

July 27th.—Morning broke on a bright bounding sea and clear blue sky. From the Testing-Room came gratifying reports of the improved insulation of

E.Walker, lith from a drawing by R. Dudley   London. Day & Son, Limited, Lith

VIEW (LOOKING AFT) FROM THE PORT PADDLE BOX OF GREAT EASTERN _ SHOWING THE TROUGH FOR CABLE &

the Cable, which had been caused by the immersion of the Cable in colder water. We were now approaching an undulation in the bed of the Atlantic in which the soundings decreased rather abruptly from 2,100 to 1,529 fathoms. The engineers were perfectly satisfied with the manner in which the machinery was working, and the mode in which the Cable ran out. The complete success of the enterprise, after this fair start, appeared to be a matter beyond doubt. The fore tank was now got ready for the paying-out of the Cable as soon as the coils in the after tank should be exhausted, and the framework for the crinoline was erected over the hatchway. At noon, our position by observation was Lat. 52° 34 30″, Long. 19° 0′ 30″, distance run 141 miles, distance from Valentia 320 miles, Cable paid out 158 miles. The Terrible was on our port beam at some distance, but the Sphinx was nowhere visible, although our speed had not much exceeded 6 knots an hour. There was in the universal benevolence of the moment a feeling of sympathy for our lagging guardians. The conviction grew that the work was nearly accomplished. Some were planning out journeys through the United States, others speculated on the probability of sport in Newfoundland : the date of our arrival was already determined upon. The sound of the piano, a tribute to our own contentment, rose from the saloon, and now and then the notes of a violin became entwined in the melodious labyrinth through which the amateur professors wandered with uncertain fingers. The artists sketched vigorously. Men stretched their legs lustily along the decks, or penetrated, with easy curiosity for the first time into the recesses of the Leviathan that bore them. None of them indeed found out the hiding-place of the ghost who haunts the ship ; but they discovered crypts under the tanks, and mean-dered and crept about the shafts and boilers of the tremendous gloominess—vast and dark as the Halls of Eblis. The ghost on board the Great Eastern, to which I have alluded, is believed to be the disembodied essence of a poor plate-riveter, who disappeared in some aperture of the nascent ship, never to be seen of mortal eye again, and who was supposed to have been riveted up by the hammers of preparation so closely that not even his spirit could escape. And so it, or he, is heard at all hours, with ghostly hammer, tap-tap-tapping on the iron walls of his prison as incessant as that cruel Raven, even through the clangour of donkey-engines and the crash of matter. There was now and then a slight indication of unsteadiness, which made one uncertain whether the wine was very strong or the Great Eastern unusually frolicsome ; but, as a matter of fact and truth, not a man aboard could imagine as he sat in the grand saloon that he was at sea at all. Every hour on board the ship increased our regard for all her qualities, except her capacity of making noise and producing smoke, but both of these were tokens and necessary conditions of her high working energies.

July 28th.—A night more of joyous progress—all going on most successfully —not a hitch in Cable, machinery, or ship. It was worth while to go aft and look at the Cable as, every inch scanned by watchful eyes, and noted in books, it flew through the whole apparatus of jockeys and drums and dynamometers, and then in a gentle curve skimmed the surface of the ocean more than 200 feet astern ere it went "plump, plunging down amid the assembly of the whales." Our course was N.W. ½ W., and the wind at W.N.W., not too strong, was just what we desired. The Terrible kept on our port beam. The Sphinx was not to be seen. Our position at noon was Lat. 52° 45′, Long. 23° 18′ 4″ (another reading gave 23° 15′ 45″), distance run since yesterday 155½ miles, Cable paid out 174 miles. Distance from Valentia 474 miles ; distance from Heart's Content 1,188·5 miles. The water was supposed to vary from 1,529 to 1950 fathoms in depth. There was something almost monotonous in our success ; no ships to be seen, only our severe-looking consort, with her black hull and two funnels and paddle-boxes, on the round blue shield of which the Great Eastern was the boss. Even the sea-birds had begun to leave us, and a whale and a few porpoises which revealed their beauties to a favoured few were regarded as an envied treat. As the departure of the Sphinx had left one flank open, and that the most vulnerable, the Great Eastern signalled to the Terrible to prevent any vessel from the N.W. crossing our course, and soon afterwards the man-of-war steamed and took up her station on our starboard quarter, where she remained throughout the day and night. There was a sense of companionship in seeing her near us.

Saturday, July 29th.—"Everything has gone on most admirably during the night." Such was the report from electricians, and engineers, and officers this morning. The electrical condition of the Cable furnished results most satisfactory to Mr. Varley and to Professor Thomson. The tests showed that in copper-resistance, insulation, and every other particular, the Cable was exhibiting an excellence far beyond the specified standard. Coil after coil whirled off from the tank and passed away to sea as easily as the lightning flash itself; and Valentia was joined to us by a lengthening thread, which seemed stronger and more sentient as it lengthened. In the night the Terrible had vanished, but she came in sight in the morning, and drew up closer to us. As the sea was calm, and the Cable ran out so beautifully, the speed of the steamer, and consequent rate of paying-out of the Cable, were increased ; and it looked as if there was really no limit to the velocity at which the process could be conducted under favouring circumstances. Yes ;

"Heart's Content" on August 5th was certain. What could prevent it? The fault which had occurred was caused by an accident most unlikely to happen again. So we pored over our maps and marked out the soundings in the little bay in Newfoundland, and imagined what sort of place it was, as men will do of spots they have never visited.

At noon our position was, Lat. 52° 33′ 30″ (another reading, 52° 38′ 30″), Long. 27° 40′. Distance run, 160 miles. Distance from Valentia, 634·4 miles. Distance to Heart's Content, 1,028 miles. The Great Eastern had passed over the valley in the plateau where the Atlantic deepens to 2,400 fathoms. At 9 a.m. we had shoaled our water to 2000 fathoms, or 2 nautical miles.

Happy is the Cable-laying that has no history. Here might the day's record have well been closed. But it was not so to be. At 1·10 p.m. (ship's time), an ill-omened activity about the Testing-Room, which had been visible for some time, reached its climax. The engines were slowed, in five minutes the great ship was motionless. In an instant afterwards every one was on deck, and the evil tidings flew from lip to lip. Something was wrong with the Cable again. But the worst was not known. "Another fault," was the word. When I went into the Testing-Room and found all the electricians so grave, I suspected more serious mischief than a diminution of insulation; and so it was. They had found "dead earth"—in other words, a complete destruction of insulation, and an uninterrupted escape of the current into the sea. About 716 miles (nautical) had been payed-out when the ship stopped so suddenly. Up to 2·40 o'clock, p.m. (Greenwich time), signals had been received from the shore in regular routine. At 3 o'clock the electricians on board began to send the current through to the shore, and in three minutes afterwards the galvanometer indicated "dead earth." So it was pretty clear the injury was close to the ship, and had gone over in the interval between 2·40 p.m. and 3·4 p.m. At 3ʰ 3′ 30″ (Greenwich time), the electrician on duty saw the index light of Thomson's galvanometer fly out of bounds whilst he was passing a current to Valentia. The nature of the injury was so decided as to admit of no doubt.

But in order to make assurance doubly sure two cuts were made in the Cable, whilst the steam was being got up forward to be in readiness for the most retrograde of all backward movements—picking-up. The whole length of Cable in the tanks was first tested, and found to be in admirable condition. Then a test outward gave "dead earth" not far overboard. The next cut at the bottom of the coil in the after tank gave the same result. The third cut was near the top of the coil in the after tank, and confirmed the testimony of the other two tests. The usual preparations were then made to shackle the Cable ere it was cut and cast overboard with its tow rope of iron wire, an operation which always caused the gravest misgivings. It was admitted that there was a certain amount of danger in it, and more in the picking-up ; but then, when the question was asked " What would you do ? " the answer was not so easy. At first it might appear natural to back the ship, and take up the Cable from the stern ; but unfortunately ships in general will not steer stern foremost, and the Great Eastern certainly would not. It was obvious that if Cables could not be secured against " faults," the mode of taking them in would have to be amended.

This was one of the most harassing days we had yet encountered ; but it proved not to be the most trying we were to endure in our short eventful history. All our calculations were falsified. Newfoundland was seen at its true distance, the piano ceased, men discussed various schemes for avoiding the transfer of the Cable from stern to the bow, on every occasion of picking-up. But all our difficulty had been overcome with such certainty, and it was so evident all would go well if no more faults existed in the Cable, that faith in the ultimate success of the enterprise became strengthened rather than diminished.

Whilst the tests were being made the Cable was running out by its own weight and the drifting of the ship, at a strain varying from 8 cwt. to 20 cwt., giving at every fathom an increase of labour in the subsequent picking up. The sailors regarded the process of cutting the Cable with distrust ; but the Cable men, accustomed to it, had no such serious apprehensions. Still the whole system of iron chains, iron rope, stoppers, and bights, is very complicated. The Cable cannot be checked in such cases till an instant before it is cut, and must be let run out for fear of the ship dragging upon it ; and to the inexperienced eye it looked as if the Great Eastern were bent on snapping the thin black thread which cut the waves like a knife-blade as she rose and fell on the swell. When the strain increased, the Cable ran with an edge of seething foam frittering before it backwards and forwards in the track of the ship, taut as a bar of steel. It was a relief to see the end cut at last, and splash over, with shackle chain and wire rope, into the water. Then began an orderly tumult of men with stoppers and guy ropes along the bulwarks and in the shrouds, and over the boats, from stern to stem, as length after length of wire rope flew out after the Cable. The men under the command of Mr. Canning were skilful in their work ; but as they clamoured and clambered along the sides, and over the boats, and round the paddle-boxes, hauling at hawsers, and slipping bights, and holding on and letting go stoppers, the sense of risk and fear for the Cable could not be got out of one's head. The chief officer, Mr. Halpin, by personal exertion, made himself conspicuous, and rendered effectual assistance ; and Capt.

Anderson, on the bridge, watched and directed every movement of the ship with skill and vigilance. But still pitches and foulings would take place for an instant, and it needed all our confidence in Mr. Canning and his staff to tolerate this picking-up system with any temper. Thousands of fathoms down we knew the end of the cable was dragging along the bottom, fiercely tugged at by the Great Eastern through its iron line. If line or Cable parted, down sank the Cable for ever. At last our minds were set at rest by the commencement of the restorative process. The head of the Great Eastern was got round slowly, and pointed eastwards. The iron wire rope was at length coming in over the bows through the picking-up machinery. In due, but in weary time, the end of the Cable appeared above the surface, and was hauled on board and passed aft towards the drum. The stern is on these occasions deserted; the clack of wheels, before so active, ceases; and the forward part of the vessel is crowded with those engaged in the work, and with those who have only to look on. The little chimneys of the boilers at the bows vomit forth clouds of smoke, the two eccentric-looking engines working the pick-up drums and wheels make as much noise as possible, brakesmen take their places, indicator and dynamometer play their parts, and all is life and bustle forwards, as with slow unequal straining the Cable is dragged up from its watery bed.

The day had been foggy or rather hazy. Light grey sheets of drizzling cloud flew over the surface of the sea, and set men talking of icebergs and Arctic storms; but towards evening the wind fell, and a cold clammy vapour settled down on ship and sea, bringing with it a leaden calm; so that the waves lost their tumbled crests, and slept at last in almost unmurmuring slumber. But the big ship slept not. The clank and beat of machinery ceased never, and the dull mill-like clatter of Cable apparatus seemed to become more active as the night wore on. The forge fires glared on her decks, and there, out in the midst of the Atlantic, anvils rang and sparks flew; and 'the spectator thought of some village far away, where the blacksmith worked, unvexed by Cable anxieties and greed of speedy news. As the blaze shot up, ruddy, mellow, and strong, and flung arms of light aloft and along the glistening decks, and then died into a red centre, masts, spars, and ropes were for the instant touched with a golden gleaming, and strange figures and faces were called out from the darkness—vanished—glinted out again—rushed suddenly into foreground of bright pictures, which faded soon away—flickered—went out—as they were called to life by its warm breath, or were buried in the outer darkness! Outside us all was obscurity; but now and then vast shadows, which moved across the arc of lighted fogbank, were projected far away by the flare; and one might well pardon the passing mariner whose bark drifted him in the night across the track of the great ship, if, crossing himself and praying with shuddering lips, he fancied he beheld a phantom ship freighted with an evil crew, and ever after told how he had seen the workshops of the Inferno floating on the bosom of the ocean. It was indeed a most wondrous and unearthly sight! The very vanes on the mastheads, the ring-bolts in the bulwarks and decks, the blocks and the cordage, were touched with such brightness that they shone as if on fire; whilst the whole of the fore part of the ship was in darkness; and on looking aft, it appeared as though the stern were on fire, or that blue lights were being burned every moment. For hour after hour, the work of "picking-up" went on. The term is objectionable; it rather indicates a brisk, lively process—a bird picks up a worm—a lady picks up a pin—a sharper picks up a flat—but the machine working at the bows of the Great Eastern assuredly was not in any one way engaged in brisk or lively work. Most doggedly at times did the Cable yield. As if it knew its home was deep in the bed of the Atlantic, and that its insulation and all the objects of its existence would be gained and bettered by remaining there, it strained against the power which sought to pull it forth; and the dynamometer showed that the resistance of the rigid cord was equivalent to $2\frac{1}{2}$ tons. At times, again, it came up merely with coy reluctance, and again became sullen as though it were already troubled by the whims of two worlds and partook of their fancies. No trace was visible of its having touched the bottom for the $2\frac{1}{2}$ miles which were hauled in, but the men observed signs of animal life on it, and certain creatures which they called "worms" were detached and fell on deck, a specimen of which I sought for in vain. As the Cable was hauled in, the men who coiled it aft, and guided it through the machinery, felt it carefully with their hands to detect any "fault" or injured part, and the line of large ship's lanterns hung up along the deck showed how carefully they did their work. It was 5·40 p.m., Greenwich time, or about 3·40 p.m., ship's time, when the end of the Cable came in board; but it was not till six hours and ten minutes had elapsed (9·50 p.m., ship's time) that the part of the Cable where the mischief lay was picked up. The defective portion was found at the very part of the Cable which was going over the stern when the ocean galvanometer indicated "dead earth." It was at once cut out, and reserved to be examined by Mr. Canning. The necessary steps were next taken to test the rest of the Cable. The shore end was spliced and jointed to a fresh end of the Cable from the after tank. These operations were finished before midnight; but it was not judged expedient to resume the process of paying-out till the morning. As yet no one knew the nature of the injury to the Cable. No one could account for the hitch; but it certainly did not affect any one's belief in success. Mr. Field, to

G. Mc Culloch lith. from a drawing by R. Dudley

London. Day & Son, Limited, Lith

THE FORGE ON DECK.-NIGHT OF AUGUST 9TH PREPARING THE IRON PLATING FOR CAPSTAN

whom such accidents are never discouraging, remarked pleasantly during the crisis of picking-up, " I have often known Cables to stop working for two hours, no one knew why, and then begin again. Most likely it's some mistake on shore." What can discourage a believer ? It was even comfort to him to remember that this very day eight years ago, a splice was made in the first Atlantic Cable, very much in the same place. But to all it had been a most trying day. And when night came, and some retired to the rest they had won so well, there, constant on the paddle-box, stood Captain Anderson, watching the course and conduct of his ship.

If the paying-out could have been stopped at once, and the Cable taken in over the stern, the delay would have been very trifling ; but that was impossible. The picking-up (necessarily slow under the most favourable circumstances) was rendered unusually tedious by the inefficiency of the boilers. An interval of 19 hours had occurred, and these faults and stoppages had caused so much labour and anxiety that Captain Anderson was obliged to remain on deck for 26 hours, whilst Mr. Halpin, Mr. Clifford, Mr. Canning, the electricians, and the whole staff, were exposed to an equal strain till the Cable was over the paying-out wheels again.

July 30th (Sunday).—The weather was exceedingly thick all night—a fog hung round the ship, and the drizzling rain was so cold as to give an impression there was ice close at hand, but the water showed it was erroneous, as the temperature was 58°. It was a dead calm, and the Great Eastern seemed to float on a grey and polished surface of cloud. The preparations for paying-out were completed and tested. There would have been a better result had not an accident occurred this morning as the Cable was being passed aft from the bow, in order to transfer it from the picking-up to the paying-out machinery. Owing to a sudden jar it flew off from the drum, and before the machinery could be stopped several fathoms had become entangled amid the wheels, and were so much injured that it was necessary to cut out the pieces, and make two new splices and joints. At 10·8 a.m. (ship's time being 8·10 a.m.) the Cable was veered out astern once more, our communications with Valentia being most satisfactory. The Cable electrically was all that could be desired, its condition being represented by 1,500,000,000 British Association units. At noon our position was Lat. 52° 30', Long. 28° 17' ; distance from Valentia, 650·6 miles ; Cable payed-out, 745 miles.

The Cable which was recovered yesterday was strained, and lay twisted in hard curves, presenting a very different appearance from the easy ductile lines in which it lay in the tank. The defective portion of the Cable was not examined to-day, and divine service was postponed till 2·30, in order to give some time for sleep and rest to the exhausted and hard-worked staff and workers of all kinds on board the ship. The weather continued thick and hazy, a fresh breeze from the N.N.W. not dispersing the cold grey clouds and mist. The Terrible alone was in sight, and it was conjectured that the Sphinx must have passed on during the night, and that she would arrive in Heart's Content before us. The sound and sight of the wheels and drums revolving again after so long a rest were very gratifying, and it was fondly hoped that this fault or dead earth would be the last, as it was now evident nothing else was to be feared, and nothing else humanly speaking could prevent the Cable being laid. In the Cable itself lay all the sources of mischief. If there were no faults or dead earth, the paying-out was a matter of the most easy routine and most positive certainty. When the operation had to be reversed, the whole condition of affairs was reversed also. A swerve of the helm, a rolling billow, an unseen weakness, a moment's neglect, the accident of an instant, and down went the thread of thought between two continents, with all which depended on it, to rest and rust in the depths of the sea. My mind could never get rid of the image of the Great Eastern pulling at the Cable as if she were animated by a malevolent desire, when she caught some one off the watch, to use her giant's strength to tear it asunder. Captain Anderson only expressed the feelings of all who watched the struggle whilst Cable and Ship were adjusting their mutual relations, when—admitting the task was more difficult than he had anticipated, in consequence of the obstacles to the management of the ship, arising from want of steerage way as soon as the engines were stopped—he said, "One feels so powerless—one can do so little to govern events while the affair of picking-up is going on." The weather was favourable, the ship perfection, and yet here were these delays arising from causes no one could foresee or prevent or remedy in any but the one way, and that a way fraught with danger. A visit to the stern, where the Cable was rolling away into 2000 fathoms water as easily as the thread flies from the reel in a lady's workbasket, always created a conviction that the enterprise must be carried out; and it was not till the machinery stopped and the words "another fault" recalled us to a sense of the contingencies on which it depended, that we could entertain a doubt of its speedy consummation. For the most indifferent somehow or another became soon interested in the undertaking. There was a wonderful sense of power in the Great Ship and in her work; it was gratifying to human pride to feel that man was mastering space, and triumphing over the winds and waves; that from his hands down in the eternal night of waters there was trailing a slender channel through which the obedient lightning would flash for ever instinct with the sympathies, passions, and interests of two mighty nations, and binding together the very ends of the earth. And then came "a fault"—or "dead earth" spoke to us.

Monday, July 31*st.*—We have been passing over the valley in the Atlantic which is more than two miles deep. With the morning came the news that all had gone well during the night. Some had got up an hour after midnight to watch the transfer of the coil from the after to the fore tank, which was looked forward to with interest, as it was supposed to be attended with some little difficulty. But they were agreeably disappointed; the operation was effected with the utmost facility. At 3·30 o'clock a.m. the ship was stopped, to permit the transfer to be made. At 3·50 a.m. the Cable was running out of the fore hold, passing down the trough, and going out over the stern as she steamed ahead again. The Great Eastern was now near a fatal spot—somewhere below us lay the bones of three Atlantic Cables.

But all during the forenoon, engineers and electricians, agreed in the most favourable statements respecting the Cable and its progress. At 9 a.m. (Greenwich time) 868 miles had been run out, and 770 miles made from land. In the forenoon Mr. Canning brought to trial the coils in which the peccant part that had wrought such mischief existed. The Court was held at the door of the Testing-Room. Mr. de Sauty acted as judge. The jury consisted of cells, wires, and galvanometers. The accused cable, cut in junks, was subjected to a silent examination, and many fathoms were pronounced not guilty, flake by flake, till at last the criminal was detected and at once carried off by Mr. Canning. The process of examination was conducted in Mr. Clifford's cabin, to which a few anxious spectators were admitted. The core was laid bare by untwisting the strands of Manilla covered with iron, and before a foot of it was uncovered an exclamation literally of horror escaped our lips! There, driven right through the centre of the coil so as to touch the inner wires, was a piece of iron wire, bright as if cut with nippers at one end and broken off short at the other. It was tried with the gauge, and found to be of the same thickness as the wire used in making the protecting cover of the Cable. On examining the strands a mark of a cut was perceived on the Manilla where the wire had entered, but it did not come through on the other side. In fact, it corresponded in length exactly with the diameter of the Cable, so that the ends did not project beyond the outer surface of the covering. Now here was at once, we thought, demonstration of a villanous design. No man who saw it could doubt that the wire had been driven in by a skilful hand. And as that was so, was it not likely that the former fault had been caused in a similar manner, and that it was not the result of accident? Then, again, it was curious that the former fault occurred when the same gang of men were at work in the tank. It was known there were enemies to the manufacturers of the Cable ; whispers went about that one of the cablemen had expressed gratification when the first fault occurred. It was a very solicitous moment, and Mr. Canning felt a great responsibility. He could not tell who was guilty, and in trying to punish them or him he might disgust the good men on whom so much depended. He at once accepted an offer made by the gentlemen on board the ship to take turn about in doing duty in the tank and superintending the men engaged in paying-out the Cable. Then he caused the cablemen to be summoned at the bows, and showed them the coil and the wire. After they had examined it curiously, he asked the men what they thought of the injury, and they one and all, without hesitation, expressed their opinion that it must have been done on purpose by some one in the tanks. Lynch law was talked of, and if the man could have been pounced upon, and left to the mercy of his fellows, he would have fared ill that day. Nor was the feeling of anger and indignation diminished by the knowledge that the punishment awarded by law for offences of such a character was a paltry fine and short imprisonment. The men who were engaged in the tank at the time of the occurrence were transferred to other duties, and the volunteer inspectors established a roster, and began their course of duty—one going on for two hours at a time, and being relieved in order, so that night and day the men engaged in paying-out the Cable were under the eyes of very vigilant watchmen. It was a painful thing to have to do, but the men admitted it was not only justifiable but necessary, and declared they were very glad the measure was adopted. It was fondly hoped that this surveillance would save us from a recurrence of the delay to which the expedition had been subjected, and ulterior steps were postponed till the shore was reached, when it was intended to institute a rigid inquiry. At noon our position was, Lat. 52° 9′ 20″, Long. 31° 53′. Length of Cable payed-out since yesterday 134 miles : total length paid out, 903 miles. Distance, from Valentia, 793 miles ; from Heart's Content, 871·9 miles. We had crossed the centre of the arc of the great circle.

Tuesday, August 1st.—The Great Eastern continued on her way without let or hindrance all night and early morning, increasing her speed to 7 knots an hour, although there was a strong breeze at times, The sea continued to favour us greatly, and the ship's deck scarcely ever varied from a horizontal plane. At noon our position was, Lat. 51° 52′ 30″, Long. 36° 3′ 30′ : making 155 miles run since yesterday. Cable paid out 1081·55 miles. Distance from Valentia, 948 miles : distance from Heart's Content, 717 miles. We were without soundings ; but it was supposed we were passing over the line on the chart where they varied from 1975 to 2250 fathoms. The Terrible was at her usual station, about two miles away ; but we gave up all hopes of seeing the Sphinx till we reached Heart's Content. It was calculated that at our present rate we would see

From a drawing by R. Dudley.

London Day & Son, Limited. Lith.

SEARCHING FOR FAULT AFTER RECOVERY OF THE CABLE FROM THE BED OF THE ATLANTIC. JULY 31ST

land on Friday evening, or first thing on Saturday morning. In preparation for our arrival the crew were employed in transferring the shore end of the Cable from the main to the after tank. It would be painful to dwell on the tenour of our conversation. The wisest men forgot the lessons of the past few days. It seemed quite certain that the right step had been taken, and that the man, or men, who had caused the previous mishaps had been effectually checkmated. The praises of the Great Eastern were on every tongue. Had no fault occurred, our task would have been nearly ended by this time. Her mission is undoubtedly the laying of Atlantic Cables, and she did it nobly as far as in her lay on this occasion.

Wednesday, August 2nd.—In the course of the night the wind, accompanied by a dense fog, rose from the westward. Then it suddenly shifted to N.N.W.; but although the sea was high, there was no rolling or pitching, and none of the sleepers were aroused from slumber, which was favoured by the ceaseless rumble of the machinery. They were, however, awakened but too speedily. Again the great enterprise on which so much depended, and on which so many hearts and eyes were fixed, was rudely checked.

As I have said, the gale did not in the least affect the ship. She went on through the heavy sea steady as an island, running out the Cable at the rate of 7 knots an hour ; and when the wind shifted to N.N.W. our course was altered to N.W. by W. $\frac{1}{2}$ W., through a sea which fell as rapidly as it had risen. The crisis was now at hand. I was aroused about 8 o'clock a.m., Greenwich time (ship's time being more than two hours earlier), by the slowing of the engines, and on looking out of my port saw, from the foam of the paddles passing ahead, that the ship was moving astern. In a moment afterwards I stood in the Testing-Room, where Mr. de Sauty, the centre of a small group of electricians, among whom was Professor Thomson, was bending over the instruments, surrounded by his anxious staff. The chronometer marked 8·6 a.m., Greenwich time. In reply to my question as to what was wrong, Professor Thomson whispered, " Another bad fault." This was indeed surprising and distressing.

In order to make the history of the day consecutive, I will relate as closely as possible what occurred. Mr. Field went on duty in the tank in the early morning, relieving M. Jules Despescher. Some twenty minutes before the fault was noticed, whilst Mr. Field was watching, a grating noise was heard in the tank as the coil flew out over the flakes. One of the men exclaimed, " There goes a piece of wire." The word was passed up through the crinoline shaft to the watcher. But he either did not hear what was said, or neglected to give any intimation, as the warning never reached Mr. Temple, who was on duty at the stern at the time. At 8 a.m., Greenwich time, being the beginning of an hour, and therefore the

L

time when in regular series the electricians on board the Great Eastern began to send currents to the shore, the gentleman engaged in watching the galvanometer, saw the unerring index light quiver for an instant and glide off the scale. The fact was established that instead of meeting with the proper resistance, and traversing the whole length of the Cable to the shore, a large portion of the stream was escaping through a breach in the gutta percha into the sea. If the quantity of the current escaping had been uniform, the electricians could calculate very nearly the distance of the spot where the injury had taken place. In the present instance, however, the tests varied greatly, and showed a varying fault. When the current is sent through a wire from one pole it produces an electro-chemical action on the wire, and at the place of the injury, which leads to a deposit of a salt of copper in the breach, and impedes the escape of electricity; and when the opposite current is returned, the deposit is reduced, and hydrogen gas formed, a globule of which may rest in the chink, and, by its non-conducting power, restore the insulation of the Cable for a time. The fault in the present instance was so grave that it was resolved to pick up the Cable once more, till we cut it out, and re-spliced it. How far away it was no one could tell precisely; but from a comparison of time it was imagined that the faulty part was not far astern, and that it was in the portion of Cable which went over at 8 o'clock in the morning, or a little before it; and although the time was not accurately fixed when Mr. Field heard it, the grating noise was supposed to arise from some cause connected with the fault. Had the engineers foreseen what subsequently occurred they might have resolved to go on, and take the chance of working through the fault. Professor Thomson has since given it as his opinion that the fault could have been worked through, and that the Cable could have transmitted messages for a long time at the rate of four words a minute— making an amply remunerative return. Mr. de Sauty also entertained the belief that the Cable could have worked for several months, at all events. But it does not appear that Mr. Canning had any reason to act on the views of these gentlemen, and it was quite sure, when the end was landed in Heart's Content, Mr. Varley could not have given his certificate that the Cable was of the contract standard. Neither Mr. Varley nor Mr. Professor Thomson had any power to interfere, or even to express their opinions, and electricians and engineers are generally inclined to regard with exclusive attention their own department in the united task, and to look to it solely.

Nothing was left but to pick up the cable. Steam was got up in the boilers for the picking-up machinery, the shackles and wire rope were prepared, and, meantime, as the ship drifted the Cable was let run out, and the brakes were regulated to reduce the strain below 30 cwt. As they were cutting the Cable near the top of the tank in the forenoon to make a test, one of the foremen perceived in the flake underneath that which had passed out with the grating noise when the fault was declared, a piece of wire projecting from the Cable, and when he took it in his fingers to prevent it catching in the passing coil, the wire broke short off. I saw it a few minutes afterwards. It was a piece of the wire of the Cable itself, not quite three inches long ; one end rather sharp, the other with a clean bright fracture, and bent very much in the same way as the piece of wire which caused the first fault. This was a very serious discovery. It gave a new turn to men's thoughts at once. After all, the Cable might carry the source of deadly mischief within itself. What we had taken for assassination might have been suicide. The piece of wire in this case was evidently bad and brittle, and had started through the Manilla in the tank. How many similar pieces might have broken without being detected or causing loss of insulation ? The marks of design in the second fault were very striking ; but the freaks of machinery in motion are extraordinary, and what looked so like purposed malice might, after all, be the effect of accidental mechanical agency. There were thenceforth for the day two parties in the ship—those who believed in malice, and those who attributed all our disasters to accident. In the end the latter school included nearly all on board the ship, and it was generally thought that in the Cable, or, rather, in what had been intended as its protection, was the source of its weakness and ruin.

Before the end of the Cable was finally shackled to the wire rope, tests were applied to the portion in tanks. The first cut was made at the old splice, between the main and fore tanks, and the Cable was found perfect. The second cut, at three miles from the end of the Cable, showed the fault to be overboard. Whilst the tests were going on, and the cablemen got the picking-up gear in readiness, the dynamometer showed a strain on the Cable astern varying from 20 to 28 cwt.

The chain and rope were at last secured to the Cable, under the eyes of Mr. Canning. It was then 9·53 a.m. The indicator stood at 376·595, showing that 1,186 miles of Cable had been payed-out. At 9·58 a.m. (Greenwich time), the Cable was cut and slipped overboard astern, fastened to its iron guardians. The depth of water was estimated at 2000 fathoms. As it went over and down in its fatal dive, one of the men said, "Away goes our talk with Valentia." Mr. de Sauty did not inform the operator at Valentia of the nature of the abrupt stoppage. We had now become so hardened to the dangers of the slip overboard, and the sight of the Cable straining for its life in contest with the Big Ship, that the cutting and slipping excited no apprehension ; but nothing could reconcile men to the picking-up machinery, and its monotonous retrogression. The wind was on our starboard beam, and the Cable was slipped over at the port quarter, and carried round on the port side towards the ship's bows, in order that the vessel might go over it, and then come up more readily to the Cable, head to wind, when the picking-up began. The drift of the ship was considerable, and it was not easy—indeed, possible—to control her movements; but, notwithstanding all this, the wire buoy-rope was got up to the machinery in reasonable time. Still the ship's head—do what Capt. Anderson would, and he did as much as any man could—did not come round easily. Even a punt will not turn if she has no way on her, and it takes a good deal of way—more than she could get with safety to the Cable—to give steerage to the Great Eastern. As she slowly drifted and came round by degrees quite imperceptible to those who did not keep a close watch on the compass, the wire rope was payed-out; and at last, as the ship's bows turned, it was taken in over the machinery, and was passed aft through the drums, and the picking-up apparatus coiled it in very slowly away till the end of the Cable was hauled up out of the sea.

It was 10·30 a.m., Greenwich time, when the Cable came in over the bow. We were now in very deep water, but had we been a few miles more to the west we should have been over the very deepest part of the Atlantic Plateau. It was believed the fault was only six miles away, and ere dead nightfall we might hope to have the fault on board, make a new splice, and proceed on our way to Heart's Content, geographically about 600 miles away. The picking-up was, as usual, exceedingly tedious, and one hour and forty-six minutes elapsed before one mile of Cable was got on board; then one of the engines' eccentric gear got out of order, and a man had to stand by with a handspike, aided by a wedge of wood and an elastic band, to aid the machinery. Next the supply of steam failed; and as soon then as the steam was got up, there was not water enough in the boiler, and so the picking-up ceased altogether. But at last all these impediments were remedied or overcome, and the operation was proceeded with before noon. Let the reader turn his face towards a window and imagine that he is standing on the bows of the Great Eastern, and then on his right will be the starboard, on his left the port side of the ship. The motion of the vessel was from right to left, and as she drifted, she tugged at the Cable from the right hand side, where he seemed to be anchored in the sea. There was not much rolling or pitching, but the set of the waves ran on her port-bow. There are in the bows of the Great Eastern two large hawse-pipes, the iron rims of which project beyond the line of the stem; against one of these the Cable caught on the left-hand side whilst the ship was drifting to the left, and soon began to chafe

From a drawing by R. Dudley.

London. Day & Son, Limited, Lith.

IN THE BOWS. AUGUST 2ND THE CABLE BROKEN AND LOST. PREPARING TO GRAPPLE..

and strain against the bow. The Great Eastern could not go astern, lest the Cable should be snapped, and without motion there was no power of steerage. At this critical moment, too, the wind shifted, so as to render it more difficult to keep the head of the ship up to the Cable. As the Cable chafed so much that there was danger of its parting, a shackle, chain, and rope belonging to one of the Cable-buoys were passed over the bows, and secured in a bight below the hawse-pipe to the Cable. These were then hauled so as to bring the Cable to the right-hand side of the bow, the ship still drifting to the left, and the oblique strain on the wires became considerable, but it was impossible to diminish it by veering out, as the length of Cable after it was cut at the stem for the operation of picking-up left little to spare. In the bow there is a large iron wheel with a deep groove in the circumference (technically called a V wheel), by the side of which is a similar but smaller wheel on the same axis. The Cable and the rope together were brought in over the bows in the groove in the larger wheel, the Cable being wound upon a drum behind by the picking-up machinery, which was once more in motion, and the rope being taken in round the capstan. But the rope and Cable did not come up in a right line in the V in the wheel, but were drawn up obliquely. Still, up they came. The strain shown on the dynamo-meter was high, but was not near the breaking point. The part of the Cable which had suffered from chafing was coming in, and the first portion of it was inboard; suddenly a jar was given to the dynamometer by a jerk, caused either by a heave of the vessel or by the shackle of wire-rope secured to the Cable, and the index jumped far above 60 cwt., the highest point marked on it. The chain shackle and wire-rope clambered up out of the groove of the V wheel, got on the rim, and rushed down with a crash on the smaller wheel, giving a severe shock to the Cable. Almost at the same moment, as the Cable and the rope travelled slowly along through the machinery, just ere they reached the dynamometer the Cable parted, flew through the stoppers, and with one bound leaped over intervening space and flashed into the sea. The shock of the instant was as sharp as the snapping of the Cable itself. No words could describe the bitterness of the disappointment. The Cable gone! gone for ever down in that fearful depth! It was enough to move one to tears; and when a man came with the piece of the end lashed still to the chain, and showed the tortured strands—the torn wires—the lacerated core—it is no exaggeration to say that a feeling of pity, as if it were some sentient creature which had been thus mutilated and dragged asunder by brutal force, moved the spectators. Captain Moriarty was just coming to the foot of the companion to put up his daily statement of the ship's position, having had excellent observations, when the news came. " I fear," he said, " we

will not feel much interested now in knowing how far we are from Heart's Content." However, it was something to know, though it was little comfort, that we had at noon run precisely 116·4 miles since yesterday; that we were 1,062·4 miles from Valentia, 606·6 miles from Heart's Content; that we were in Lat. 51° 25', Long. 39° 6', our course being 76° S. and 25° W. But instant strenuous action was demanded! Alas! action! There around us lay the placid Atlantic smiling in the sun, and not a dimple to show where lay so many hopes buried. The Terrible was signalled to, "the Cable has parted," and soon bore down to us, and came-to off our port beam. After brief consideration, Mr. Canning resolved to make an attempt to recover the Cable. Never, we thought, had alchemist less chance of finding a gold button in the dross from which he was seeking aurum potabile, or philosopher's stone. But, then, what would they say in England, if not even an attempt, however desperate, were made? There were men on board who had picked up Cables from the Mediterranean 700 fathoms down. The weather was beautiful, but we had no soundings, and the depth was matter of conjecture; still it was settled that the Great Eastern should steam to windward and eastward of the position in which she was when the Cable went down, lower a grapnel, and drift down across the course of the track in which the Cable was supposed to be lying. Although all utterance of hope was suppressed, and no word of confidence escaped the lips, the mocking shadows of both were treasured in some quiet nook of the fancy. The doctrine of chances could not touch such a contingency as we had to speculate upon. The ship stood away some 13 or 14 miles from the spot where the accident occurred, and there lay-to in smooth water, with the Terrible in company. The grapnel, two five-armed anchors, with flukes sharply curved and tapering to an oblique tooth-like end—the hooks with which the giant Despair was going to fish from the Great Eastern for a take worth, with all its belongings, more than a million, were brought up to the bows. One of these, weighing 3 cwt., shackled and secured to wire buoy rope, of which there were five miles on board, with a breaking strain calculated at 10 tons, was thrown over at 3·20, ship's time, and "whistled thro'" the sea, a prey to fortune. At first the iron sank slowly, but soon the momentum of descent increased, so as to lay great stress on the picking-up machinery, which was rendered available to lowering the novel messenger with warrant of search for the fugitive hidden in mysterious caverns beneath. Length flew after length over cog-wheel and drum till the iron, warming with work, heated so as to convert the water thrown upon the machinery into clouds of steam. The time passed heavily. The electricians' room was closed; all their subtle apparatus stood functionless, and cell, zinc, and copper threw off superfluous currents in the darkened chamber. The jockeys

E. Walker lith. from a drawing by R. Dudley

London Day & Son, Limited, Lith.

GETTING OUT ONE OF THE LARGE BUOYS FOR LAUNCHING. AUGUST 2ND.

had run their race, and reposed in their iron saddles. The drums beat no more, their long réveillée ended in the muffled roll of death ; that which had been broken could give no trouble to break, and man shunned the region where all these mute witnesses were testifying to the vanity of human wishes. All life died out in the vessel, and no noise was heard except the dull grating of the wire-rope over the wheels at the bows. The most apathetic would have thought the rumble of the Cable the most grateful music in the world.

Away slipped the wire strands, shackle after shackle : ocean was indeed insatiable ; "more" and "more," cried the daughter of horse-leech from the black night of waters, and still the rope descended. One thousand fathoms—fifteen hundred fathoms—two thousand fathoms—hundreds again mounting up—till at last, at 5·6 p.m., the strain was diminished, and at 2,500 fathoms, or 15,000 feet, the grapnel reached the bed of the Atlantic, and set to its task of finding and holding the Cable. Where *that* lay was of course beyond human knowledge ; but as the ship drifted down across its course, there was just a sort of head-shaking surmise that the grapnel might catch it, that the ship might feel it, that the iron-rope might be brought up again—and that the Cable across it might—here was the most hazardous hitch of all—might come up without breaking. But 2,500 fathoms ! Alas !—and so in the darkness of the night—not more gloomy than her errand—the Great Eastern, having cleared away one of the great buoys and got it over her bows, was left as a sport to the wind, and drifted, at the rate of 70 feet a minute, down upon the imaginary line where the Cable had sunk to useless rest.

August 3rd.—All through the night's darkness the Great Eastern groped along the bottom with the grapnel as the wind drifted her, but cunning hands had placed the ship so that her course lay right athwart the line for which she was fishing. There were many on board who believed the grapnel would not catch anything but a rock, and that if it caught a rock or anything else it would break itself or the line without anyone on board being the wiser for it. Others contended the Cable would be torn asunder by the grapnel. Others calculated the force required to draw up two miles and a-half of the Cable to the surface, and to drag along the bottom the length of line needed to give a bight to the Cable caught in the grapnel, so as to permit it to mount two and a-half miles to the deck of the Great Eastern. After the grapnel touched the bottom, which was at 7·45 o'clock, p.m., last night, when 2,500 fathoms of rope were payed-out, the strain for an hour and a-half did not exceed 55 cwt. ; but at 10 p.m. it rose to 80 cwt. for a short time, and the head of the ship yielded a little from its course and came up to the wind. It then fell off as the strain was reduced to 55 cwt.

which apparently was the normal force put on the ship by the weight of the rope and grapnel. This morning the same strain was shown by the dynamometer, and it varied very slightly from midnight till 6 o'clock a.m. Then the bow of the ship and the index of the dynamometer coincided in their testimony, and whilst the Great Eastern swayed gradually and turned her head towards the wind, the index of the machine recorded an increasing pressure. It began to be seen that there was some agency working to alter the course of the ship, and the dynamometer showed a strain of 70 cwt. The news soon spread ; men rushed from compass to dynamometer. "We have caught it ! we have caught it !" was heard from every lip.

There was in this little world of ours as much ever-varying excitement, as much elation and depression, as if it were a focus into which converged the joys and sorrows of humanity. When the Great Eastern first became sensible of the stress brought upon her by the grappling iron and rope she shook her head, and kept on her course, disappointing the hopes of those who were watching the dynamometer, and who saw with delight the rising strain. This happened several times. It was for a long time doubtful whether the grapnel held to anything more tenacious than the ooze, which for a moment arrested its progress and then gave way with a jerk as the ship drifted ; but in the early morning, the long steady pull made it evident the curved prongs had laid their grip on a solid body, which yielded slowly to the pressure of the vessel as she went to leeward, but at the same time resisted so forcibly as to slew round her bow. The scientific men calculated the force exercised by grapnel and rope alone to be far less than that now shown on the dynamometer. And if the Great Eastern had indeed got hold of a substance in the bottom of the Atlantic at once so tenacious and so yielding, what could it be but the lost Cable ?

At 6·40 a.m., Greenwich time, the bow of the ship was brought up to the grapnel line. The machinery was set to work to pull up the 2,500 fathoms of rope. The index of the dynamometer, immediately on the first revolutions of the wheels and drums, rose to 85 cwt. The operation was of course exceedingly tedious, and its difficulty was increased by the nature of the rope, which was not made in a continuous piece, but in lengths of 100 fathoms each, secured by shackles and swivels of large size, and presumably of proportionate strength. It was watched with intense interest. The bows were crowded, in spite of the danger to which the spectators were exposed by the snapping of the wire-rope, which might have caused them serious and fatal injuries. At 7·15 o'clock, a.m., the first 100 fathoms of rope were in, and the great iron shackle and swivel at the end of the length were regarded with some feelings of triumph. At 7·55 a.m. the second

T.Picken, lith.from a drawing by R.Dudley

London Day & Son,Limited,Lith.

GENERAL VIEW OF PORT MAGEE, &c. FROM THE HEIGHTS BELOW CORA BEG. THE CAROLINE LAYING THE SHORE END OF THE CABLE JULY 22ᴺᴰ.

length of 100 fathoms was on board, the strain varying from 65 to 75 cwt. At 8·10 a.m., when 400 fathoms had been purchased in and coiled away, the driving spur-wheel of the machinery broke, and the rope snapped, the strain being 90 cwt. at the time. The whole of the two miles of wire rope, grapnel and all, would have been lost, but that the stoppers caught the shackle at the end, and saved the experiment from a fatal termination. The operation was suspended for a short time, in order to permit the damage to be made good, and the rope was transferred to the capstan. The hazardous nature of the work, owing to the straining and jerking of the wire rope, was painfully evinced by the occurrence of accidents to two of the best men on Mr. Canning's staff—one of whom was cut on the face, and the other had his jaw laid open. At noon nearly half a mile of rope was gathered in. With every length of Cable drawn up from the sea, the spirits of all on board became lighter, and whilst we all talked of the uncertainty of such an accomplishment, there was a sentiment stronger than any one would care to avow, inspiring the secret confidence that, having caught the Cable in this extraordinary manner, we should get it up at last, and end our strange eventful history by a triumphant entry to Heart's Content. Already there were divers theories started as to the best way of getting the Cable on board, for if Mr. Canning ever saw the bight, the obvious question arose, " What will he do with it ? " The whole of our speculations were abruptly terminated at 2·50 o'clock, p.m. As the shackle and swivel of the eleventh length of rope, which would have made a mile on board, were passing the machinery, the head of the swivel pin was wrung off by the strain, and the 1,400 fathoms of line, with grapnel attached, rushed down again to the bottom of the Atlantic, carrying with it the bight of Cable. The shock was bitter and sharp. The nature of the mishap was quite unforeseen. The engineers had calculated that the wire rope might part, or that the Cable itself might break at the bight, but no one had thought of the stout iron shackles and swivels yielding. To add to the gloominess of the situation, the fog, which had so long been hanging round the ship, settled down densely, and obliged the Great Eastern to proceed with extreme caution. But although the event damped, it did not extinguish, the hopes of the engineers. Mr. Canning and Mr. Clifford at once set their staff to bend 2,500 fathoms of spare wire rope to another grapnel, and to prepare a buoy to mark the spot as nearly as could be guessed where the rope had parted, and gone down with the bight of the Cable. The Great Eastern was to steam away to windward of the course of the Cable, and then drift down upon it about three miles west of the place where the accident occurred. Fog whistles were blown to warn the Terrible of our change of position, and at 1·30, ship's

M

time, the Great Eastern, as she steamed slowly away, fired a gun, to which a real or fancied response was heard soon afterwards. As she went ahead, guns were fired every 20 minutes, and the steam-whistles were kept going, but no reply was made, and she proceeded on her course alone. It was impossible to obtain a noon-day observation, and the only course to be pursued was to steam to windward for 14 or 15 miles, then to lay-to and drift, in the hope of procuring a favourable position for letting go the second grapnel, and catching the Cable once more.

August 4th.—The morning found the Great Eastern drifting in a dense fog. In order to gauge the nature of the task before them, the engineers fitted up a sounding tackle of all the spare line they could get, and hove it overboard with a heavy lead attached. The sinker, it is believed, touched bottom at 2,300 fathoms, but it never came up to tell the tale. The line broke when the men were pulling it in, and 2000 fathoms of cord were added to the maze of Cable and wire rope with which the bed of the Atlantic must be vexed hereabouts. The fog cleared away in the morning, and the Terrible was visible astern. Presently one of her boats put off, with a two-mile pull before her, for the Great Eastern. Lieutenant Prowse was sent to know what we had been doing, and what we intended to do. He returned to his ship with the information that Mr. Canning, full of determination, if not of hope, would renew his attempt to grapple the Cable, and haul it up once more. At noon, Captain Anderson and Staff-Commander Moriarty, who had been very much perplexed at the obstinate refusal of the sun to shine, and might be seen any time between 8 a.m. and noon parading the bridge sextant in hand, taking sights at space, succeeded in obtaining an observation, which gave our position Lat. 51° 34′ 30″, Long. 37° 54′. The Great Eastern had drifted 34 miles from the place where the Cable parted, and as she had steamed 12 miles, her position was 46 miles to the east of the end of the Cable.

Meantime the engineers' staff were busy making a solid strong raft of timber balks, 8 feet square, to serve as a base to a buoy to be anchored in 2,500 fathoms, as near as possible to the course of the Cable, and some miles to the westward of the place where the grapnel-rope parted. A portion of Cable, which had been a good deal strained, was used as tackle, for the purpose of securing the raft and buoy to a mushroom anchor. The buoy, which we shall call No. 1, was painted red, and was surmounted by a black ball, above which rose a staff, bearing a red flag. It was securely lashed on the raft. At 10 p.m., Greenwich time, the buoy No. 1 was hove overboard, and sailed away over the grey leaden water till it was brought up by the anchor in Lat. 51° 28′, Long. 38° 42′ 30″. The Great Eastern, having thus marked a spot on the ocean, proceeded on her cruise, to take up a position which might enable her to cross the Cable with the new grapnel, and try fortune once more. Some researches made among the coils of telegraph Cable confirmed the opinion, that the iron wires in the outer protective coating were the sources of all our calamities, and fortified the position of those who maintained that the faults were the result of accident. In some instances the wires were started; in others they were broken in the strands. By twisting the wire, great variations in quality became apparent. Some portions were very tough, others snapped like steel. It is to be regretted that the scientific council who recommended the Cable did not test some parts of it in the paying-out apparatus with a severe strain, as they might have detected the inherent faults in the fabric. It is quite possible hundreds of broken ends exist in the Cable already laid, though they have done no harm to the insulation.

Saturday, August 5th.—There was no change in the weather. A grey mist enveloped the Great Eastern from stem to stern, blanket-like as sleep itself. The haze—for so it was rather than a fog—got lighter soon after 12 o'clock, but it was quite out of the question to attempt an observation of a longitudinal character. The steam-whistles pierced the fog-banks miles away. Shoals of grampuses, black fish, porpoises, came out of the obscure to investigate the source of such dread clamour, and blew, spouted, and rolled on the tops of the smooth unctuous-looking folds of water that undulated in broad sweeping billows on our beam. Our great object was to get sight of the buoy, and by that means make a guess at our position. At 12·30 p.m. the Terrible was sighted on the port beam, and our fog music was hushed. At 2·30 o'clock, p.m., the Terrible signalled that the buoy was three miles distant from her. This was quite an agreeable incident. Every eye was strained in search of the missing buoy, and at last the small red flag at the top of the staff was made out on the horizon. At 3·45 o'clock, p.m., the Great Eastern was abreast of the buoy, which was hailed with much satisfaction. It bore itself bravely, though rather more depressed than we had anticipated, and it was like meeting an old friend, to see it bobbing at us up and down in the ocean. It was resolved to steer N.W. by N. for 5 or 6 miles, so as to pass some miles beyond the Cable, and then, if the wind answered, to drift down and grapple. The Great Eastern signalled to the Terrible, " Please watch the buoy;" and, under her trusty watch and ward, we left the sole mark of the expedition fixed on the surface of the sea, and stood towards the northward. The wind, however, did not answer, and the grapnel was not thrown overboard.

Aug. 6th, Sunday.—It was very thick all through the night—fog, rain, drizzle alternately, and all together. When morning broke, the Terrible was

M 2

visible for a moment in a lift of the veil of grey vapour which hung down from the sky on the face of the waters. The buoy was of course quite lost to view, nor did we see it all day. At 10·45 a.m. Captain Anderson read prayers in the saloon. At noon it was quite hopeless to form a conjecture respecting the position of the sun or of the horizon, but Captain Moriarty and Captain Anderson were ready to pounce upon either, and as the least gleam of light came forth, sextants in hand, like the figures which indicate fine weather in the German hygrometers. The sea was calm, rolling in lazy folds under the ship, which scarcely condescended to notice them. She is a wonder! In default of anything else, it was something to lie on a sofa in the ladies' saloon, and try to think you really were on the bosom of the Atlantic,— not a bulkhead creaking, not a lamp moving, not a glass jingling. Under the influence of an unknown current, the Great Eastern was drifting steadily against the wind. When the circumstance was noticed, it could only be referred to the " Gulf Stream," which is held answerable for a good many things all over the world. At 4 p.m. the buoy was supposed to be 15 miles N.W. ½ N. of us, the wind being E.S.E., but it was only out of many calculations Captain Moriarty and Captain Anderson created a hypothetical position. There had been no good observation for three days, and until we could determine the ship's position exactly, and get a good wind to drift down on the Cable, it would be quite useless to put down the grapnel.

The buoy was supposed to be some 12 miles distant from the end of the Cable, and not far from the slack made by the Great Eastern. If we got this slack, the Cable would come up more easily on the grapnel. Of course, if the buoy had been ready when the Cable broke, it would have been cast loose at the spot where the wire rope and grapnel sank. If the Cable could be caught, it was proposed either to place a breaking strain upon it, so as to get a loose end and a portion of slack, and then to grapple for it a second time within a mile or so of the end, or to try and take it inboard without breaking. Some suggested that the Great Eastern should steam at once to Trinity Bay, where the fleet was lying, and ask the admiral for a couple of men-of-war to help us in grappling; but those acquainted with our naval resources declared that it would be useless, as the ships would have no tackle aboard fit for the work, and could not get it even at Halifax. Others recommended an immediate return to England for a similar purpose, to get a complete outfit for grappling before the season was advanced, and to return to the end of the Cable, or to a spot 100 miles east of it, where the water is not so deep. What was positive was, that more than 1,100 miles of the most perfect Cable ever laid, as regards electrical conditions, was now lying three-quarters of the way across from Valentia to Newfoundland.

Monday, Aug. 7th.—During the night it was raining, fogging, drizzling, clouding over and under, doing anything but blowing, and of course as we drifted hither and thither,—the largest float that currents and waves ever toyed with,—we had no notion of any particular value of our whereabouts. But at 4 a.m. a glimpse was caught of the Terrible lying-to about 6 miles distant, and we steered gently towards her and found that she was keeping watch over the buoy, which was floating apparently 2 miles away from her. Our course was W.N.W. till we came nearly abreast of the buoy shortly before 9 a.m., when it was altered to N.W. The wind was light and from the northward, and the Great Eastern steamed quietly onwards that she might heave over the grapnel and drift down on the line of the Cable when the fog cleared and the wind favoured.

The feat of seamanship which was accomplished, and the work so nearly consummated, was so marvellous as to render its abrupt and profitless termination all the more bitter. The remarkable difficulty of such a task as Staff-Commander Moriarty and Captain Anderson executed cannot be understood without some sort of appreciation of the obstacles before them. The Atlantic Cable, as we sadly remember, dropped into the unknown abyss on Aug. 2. We had no soundings. In the night the Great Eastern drifted and steamed 25 miles from the end of the Cable—then bore away with a grapnel overboard, and 2,500 fathoms of wire rope attached, and steered so as to come across the course of the Cable at the bottom. On the morning of Aug. 3rd, the increasing strain on the line which towed the grapnel gave rise to hope at first, and finally to the certainty, that the ship had caught the Cable. At 3·20 o'clock, p.m., Greenwich time, when about 900 fathoms of grapnel line had been hauled in, the head of a swivel pin broke, and 1,400 fathoms of line, with grapnels and Atlantic Cable, went down to the bottom. Then the Great Eastern drifted again in a fog whilst preparing for another trial to drag the Cable up from the sea, and on 4th August, with an apparatus devised on board, got doubtful soundings, from which it was estimated that the water was about 2½ miles deep. A buoy placed on a raft, which sunk so deep that only a small flagstaff and black bulb were visible, was let go, with a mushroom anchor and 2½ miles of Cable attached to it, into this profound; but as it was not ready when the Cable broke, the buoy was slipped over at the distance of some miles from the place where the fatal fracture took place, in the hope and belief that the anchor would come up somewhere near the slack caused by the picking-up operations. Still in fog, which shut the Terrible out of sight, the Great Eastern prepared for another attempt. Next day

(August 5), with the assistance of the Terrible, she came upon the buoy, and having steamed away to a favourable position, so as to come down on the course of the Cable again, remained drifting and steaming gently, on the look-out for the buoy, which it was very difficult to discover owing to the fog and to the current and winds acting on the ship. The weather did not permit any observations for longitude to be made during the whole of this period. On Aug. 7th we passed the buoy and steered N.W., and at 11·10 a.m., ship's time, 1·47 p.m., Greenwich time, another grapnel, with 2,500 fathoms of wire rope, was thrown over, and the Great Eastern, with a favourable wind, was let drift down on the course of the Cable, about half way between the buoy and the broken end. At 12·5 ship's time, the grapnel touched the bottom in 2,500 fathoms water, having sunk, owing to improved apparatus, in half the time consumed in the first operation. In six hours afterwards, the eyes which were watching every motion of the ship so anxiously, perceived the slightest possible indication that the grapnel was holding on at the bottom, and that the ship's head was coming up towards the northward. It is not possible to describe the joyous excitement which diffused itself over the Great Eastern as, with slowly-increasing certitude, she yielded to the strain from the grapnel and its prize, and in an hour and a-half canted her head from E. by S. ½ S., to E. ¾ North. The screw was used to bring up her bow to the strain, and the machinery of the picking-up apparatus, much improved and strengthened, was set in motion to draw in the grapnel by means of the capstan and its steam power. The strain shown by the indicator increased from 48 cwt. to 66 cwt. in a short time ; but the engines did their work steadily till 8·10, when one of the wheels was broken by a jerk, which caused a slight delay. The grapnel-rope was, however, hauled in by the capstan at a uniform rate of 100 fathoms in 40 minutes ; but the strain went on gradually increasing till it reached 70 cwt. to 75 cwt. At 11·30 p.m., ship's time, or 2·5 a.m., Greenwich, 300 fathoms were aboard, and at midnight all those who were not engaged on duty connected with the operation retired to rest, thankful and encouraged. In the words of our signal to the Terrible, all was going on " hopefully." Throughout our slumbers the clank of the machinery, the shrill whistles to go on ahead, or turn astern, sounded till morning came, and when one by one the citizens of our little world turned up on deck, each felt, as he saw the wheels revolving and the wire rope uncoiling from the drums, that he was assisting at an attempt of singular audacity and success. A moonlight of great brightness, a night of quiet loveliness had favoured the enterprise, and the links of rope had come in one after another at a speed which furnished grounds for hope that if the end of the day witnessed similar progress, the Cable would be at the surface before nightfall.

J. Mc Culloch lith. from a drawing by R.Dudley.

London. Day & Son, Limited. Lith.

INTERIOR. OF ONE OF THE TANKS ON BOARD THE GREAT EASTERN _ CABLE PASSING OUT.

August 8*th*.—This morning, about 7·30, one mile—one thousand fathoms—had been recovered, and was coiled on deck. The Cable, however, put out a little more vigour in its resistance, and the strain went up to 80 cwt., having touched 90 cwt. once or twice previously. No matter what happened, the perseverance of the engineers and seamen had been so far rewarded by a very extraordinary result. They had caught up a thin Cable from a depth of 2,500 fathoms, and had hauled it up through a mile of water. They were hauling at it still, and all might be recovered. But it was not so to be. Our speculations were summarily disposed of—our hopes sent to rest in the Atlantic. Shortly before 8 o'clock, an iron shackle and swivel at the end of a length of wire rope came over the bow, passed over the drums, and had been wound three times round the capstan, when the head of the swivel bolt "drew," exactly as the swivel before it had done, and the rope, parting at once, flew round the capstan, over the drums, through the stops, with the irresistible force on it of a strain, indicated at the time or a little previously, of 90 cwt. It is wonderful no one was hurt. The end of the rope flourished its iron fist in the air, and struck out with it right and left, as though it were animated by a desire to destroy those who might arrest its progress. It passed through the line of cablemen with an impatient sweep, dashed at one man's head, was only balked by his sudden stoop, and menacing from side to side the men at the bow, who fortunately were few in number, and were warned of the danger of their position, splashed overboard. All had been done that the means at the disposal of engineers and officers allowed. The machinery had been altered, improved, tested—every shackle and swivel had been separately examined, and several which looked faulty had been knocked off and replaced, but in every instance the metal was found to be of superior quality. It was 7·43 a.m., ship's time, exactly, when the rope parted. The sad news was signalled to the Terrible, which had been following our progress anxiously and hopefully during the night. Her flags in return soon said, " Very sorry," and she steamed towards the Great Eastern immediately. Mr. Canning and Mr. Gooch, and others, consulted what was best to be done, and meantime the buoy and raft which had been prepared in anticipation of such a catastrophe as had occurred, were lowered over the bows with a mooring rope of 2,500 fathoms long, attached to a broken spur-wheel. The buoy was surmounted by a rod with a black ball at the top over a flag red, white, and red, in three alternate horizontal stripes, and on it were the words and letters :—" Telegraph, No. 3." It floated rather low on a strong raft of timber, with corks lashed at the corners, and by observation and reckoning it was lowered in Lat. 51° 25′ 30″, Long. 38° 56′. The old buoy at the time it was slipped bore S.E. by E. 13 miles from the Great Eastern. As there were still nearly 1,900 fathoms of wire rope on board, and some 500 fathoms of Manilla hawser, Mr. Canning resolved to make a third and last attempt ere he returned to Sheerness. Captain Anderson warned Mr. Canning that from the indications of the weather, it was not likely he could renew his search for two or three days, but that was of the less consequence, inasmuch as it needed nearly that time for Mr. Canning's men to secure the shackles and prepare the apparatus for the third trial.

At 9·40 a.m., just as the buoy had gone over, a boat came alongside from the Terrible, and Mr. Prowse, the First Lieutenant, boarded us to know what we were going to do, to compare latitude and longitude, and to report to Captain Napier the decision arrived at by the gentlemen connected with the management of the Expedition. The Great Eastern had still about 3,500 tons of coal remaining, and the Terrible could wait three days more, and still keep coal enough to enable her to reach St. John's. At 11·30 the Great Eastern stood down to the second buoy, for the purpose of fixing its exact locality by observation. Soon afterwards the weather grew threatening, and at 2 p.m. we were obliged to put her head to the sea, which gradually increased till the Great Eastern began for the first time to give signs and tokens that she was not a fixture. The Terrible stood on ahead on our port side, and for some time we kept the buoy equi-distant between us. At night, the wind increased to half a gale, and it was agreed on all sides that though the Great Eastern could have paid out the Cable with the utmost ease, she could not have picked up, and certainly could not have kept the grapnel line and Cable under her bows in such weather. But the steadiness of the vessel was the constant theme of praise. During the night she just kept her head to the sea. The Terrible, which got on our port and then on our starboard bow, signalled to us not to come too close, and before midnight her lights were invisible on our port quarter—one funnel down.

Aug. 9th.—Our course was W.N.W. during the night; weather thick and rainy —strong southerly wind; sea running moderately high. At 6 a.m., having run by reckoning 35 miles from the buoy, our course was altered to E.S.E., so as to bring us back to it. The state of the weather delayed the artificers in their work. It rained heavily, the deck was by no means a horizontal plane, and it was doubtful if Mr. Canning and Mr. Clifford, using all possible diligence, could get tackle and machinery in order before the following forenoon, so that it was not necessary to make any great speed. The reputation of the ship was enhanced in the eyes and feelings of her passengers by the manner in which she had behaved in the undoubtedly high breeze and heavy sea. The former was admitted by sailors to be a " gale," though they seemed to think the force of the wind was affected by

E. Walker, lith. from a drawing by R. Dudley.

London. Day & Son, Limited. Lith

LAUNCHING BUOY ON AUGUST 8TH IN LAT 51° 25' 30" LONG. 38° 56' (MARKING SPOT WHERE CABLE HAD BEEN GRAPPLED)

the addition of the prefix "summer," as if it mattered much at what time of the year a gale blows. The latter, when we turned tail and went before it, soon developed a latent tendency in the Great Eastern to obey the rules governing bodies floating on liquids under the action of summer gales. She rolled with a gravity and grandeur becoming so large a ship once in every 11 or 12 seconds; but on descending from the high decks to the saloon, one found no difficulty in walking along from end to end of it without gratuitous balancings or unpremeditated halts and progresses. It was a grey, gloomy, cloudy sea and sky—not a sail or a bird visible. In the forenoon the Terrible came in sight, lying-to with her topsail set, and it was hoped she was somewhere near the buoy. At noon our position was ascertained by observation to be Lat. 51° 29′ 30″, Long. 39° 6′ 0″. Great Eastern, as soon as she was near enough, asked the Terrible, "Do you see the buoy?" After a time, the answer flew out, "No." Then she added that she was "waiting for her position," and that she "believes the buoy to be S.S.E." of us. Our course was altered S. by E. ½ E., and the look-out men in the top swept the sea on all sides. The Terrible also started on the search. At 3·20 p.m. the two ships were within signalling distance again—sea decreasing, wind falling fast. The Terrible asked, "Did you see buoy?" which was answered in negative, and then inquired if the Great Eastern was going to grapple again, which was replied to in the affirmative—Captain Anderson busy in one cabin and Staff-Commander Moriarty busy in another, working diagrams and calculations, and coming nearer and nearer to the little speck which fancies it is hidden in the ocean: with very good reason, too, for the search after such an object on such a field as the Atlantic, ruffled by a gale of wind, might well be esteemed of very doubtful success. But the merchant captain and the naval staff-commander were not men to be beaten, and in keen friendly competition ran a race with pencils and charts to see who could determine the ship's position with the greatest accuracy, being rarely a mile apart from each other in the result. The only dubious point related to the buoy itself, for it might have drifted in the gale, it might have gone down at its moorings, or the Cable might have parted. There were strong currents, as well as winds and waves. The moment the weather moderated in the forenoon, the whole body of smiths and carpenters, and workers in iron, metal, and wood, were set to work at the alterations in the machinery for letting out the grapnel and taking it in again. A little army of skilled mechanics were exercising on deck; workshops and forges were established, and some of the many chimneys which rise above the bulwarks of the Great Eastern, and put one in mind of the roofs of the streets seen from the railway approaches to London, began to smoke. The smiths forged new pins for the swivels, and made new shackles and swivels; the carpenters made casings for capstan ; ropemakers examined and secured the lengths of wire rope, and a new hawser was bent on to make up for the deficiency of buoy rope. At last, the much-sought-for object was discovered—the buoy was visible some 2 miles distant. The Great Eastern made haste to announce the news to the Terrible, and just as her flags were going aloft, a fluttering of bunting was visible in the rigging of the Terrible, and the signalman read her brief statement that the buoy was where we saw it was, thus proving that both vessels dropped on it at the same time. The finding of the little black point on the face of the Atlantic was a feat of navigation which gave great satisfaction to the worthy performers and the spectators. A little before 5 o'clock the Great Eastern was abreast of the buoy. The Terrible came up on the other side of it, and the Great Eastern and the man-of-war lay-to watching the tiny black ball, which bobbed up and down on the Atlantic swell, intending to stay by it as closely as possible till morning. By dint of energetic exertion, Mr. Canning hoped to have his grapnel and tackle quite ready the moment the ship was in position on the morrow. It was a sight to behold the deck at night—bare-armed Vulcans wielding the sledge—Brontes, Steropes, and Pyracmon at bellows, forge, and anvil—fires blazing—hailing sparks flashing along the decks—incandescent masses of iron growing into shape under the fierce blows—amateurs and artists admiring—the sea keeping watch and ward outside, and the hum of voices from its myriad of sentry waves rising above the clank of hammers which were closing the rivets up of the mail in which we were to do battle with old ocean for the captive he holds in his dismal dungeons below. Will he yield up his prisoner ?

Aug. 10*th.* A more lovely morning could not be desired—sea, wind, position —all were auspicious for the renewed attempt, which must also be the last if our tackle break. A light breeze from the west succeeded to the gale, and a strong current setting to the eastward prevailed over it, and carried the Great Eastern nearly 7 miles dead against the wind from 9 p.m. last night till 4 a.m. this morning, thus taking her away from the buoy. The swell subsided, and such wind as there was favoured the plan to drift across the course of the Cable about a mile to westward of the place where the last grapnel was lost. Without much trouble the Great Eastern, having come upon the first buoy, caught the second buoy, and both were in sight at the same moment. Authorities differed concerning their distance. One maintained they were $7\frac{1}{2}$ miles, the other that they were 10 miles apart. At 10·30, Greenwich time, when we were between $1\frac{1}{2}$ and $1\frac{3}{4}$ mile distant from the course of the Cable, the buoy bearing S.S.E., the grapnel was thrown over, and 2,460 fathoms of wire rope and hawser were paid out in 48 minutes.

As there was a current still setting against the easterly wind, which had increased in strength, Captain Anderson at first got all fore-and-aft canvas on the ship, to which were added afterwards her fore and maintopsails; her course was set N.W. by N., but she made little headway, and drifted to S.W. At 11·10 a.m., ship's time, an increased strain on the grapnel line was shown by the dynamometer, and at the same time the head of the Great Eastern began to turn slowly northwards from her true course.

The square-sails were at once taken in. Great animation prevailed at the prospect of a third grapple with the Cable. But in a few moments the hope proved delusive, and the ship continued to drift to S. and W., the buoy bearing S.E. The bow swept round, varying from W. and by N. to N. W. and by N. At noon the Great Eastern, if all reckonings were right, was but half a mile from the Cable, and the officers hoped she would come across it about half a mile west of the spot where she last hooked it. But at 3·30 p.m. the last hope vanished. The ship must by that time have long passed the course of the Cable. Captain Anderson had an idea that we grappled it for a moment soon after noon, when the ship's head came 3 points to the N., and the strain increased for a moment to 60 cwt. The buoy was now $2\frac{1}{2}$ to 3 miles E.—ship's head being W.N.W. All that could be done was to take up grapnel, and make another cast for the Cable. The wind increased from eastward. At 4·15 p.m. ship's head was set N. by E. by screw, in order to enable the grapnel line to be taken in, and the capstan was set to haul up the grapnel. The wire rope came over the bows unstranded, and in very bad condition. Much controversy arose respecting the cause of this mischief. Some, the practical men, maintaining it was because there were not swivels enough on it; others, the theoretical men, demonstrating that the swivels had nothing to do with the torsion or detorsion; and both arguing as keenly with respect to what was happening 2 miles below them in the sea as if they were on the spot. The process of pulling up such a length of wire is tedious, and although no one had expressed much confidence in the experiment, every one was chagrined at the aspect of the tortured wire as it came curling and twisting inboard from its abortive mission. At midnight 1000 fathoms had been hauled in.

August 11*th.*—Nothing to record of the night and early morning, save that both were fine, and that the capstan took in the iron fishing-line easily till 5·20 a.m., ship's time, when the grapnel came up to the bows. The cause of the failure was at once explained: the grapnel could not have caught the Cable, because in going down, or in dragging at the bottom, the chain of the shank had caught round one of the flukes. From the condition of the rope it was calculated that we were in only 1,950 fathoms of water, for nearly 500 fathoms of it were covered with the grey ooze of the bottom. The collectors scraped away at the precious gathering all the morning, and for a time forgot their sorrows.

It was now a dead calm, and Mr. Canning mustered his forces for another attempt for the Cable! He overhauled the wire rope, and exorcised hawsers out of crypts all over the ship.

"Hope lives eternal in the human breast."

Although the previous trials, with better gear, had proved unsuccessful; although the tackle now used was a thing of shreds and patches; although Mr. Canning and others said, "We are going to make this attempt because it is our duty to exhaust every means in our power," and thereby implied they had little or no confidence of success; there was scarcely a man in the ship who did not think "there is just a chance," and who would not have made the endeavour had the matter been left to his own decision. It was some encouragement to ascertain that there were only 1,950 fathoms of water below us. It was argued that, if the Cable could be broken at the bight, another drift about a mile from the loose end would be certain to succeed, as the loose end would twist round the eastward portion of the Cable, and come up at a diminished strain to the surface. A grapnel with a shorter shank was selected for the next trial. The cablemen were set to work to coil down the new rope and hawsers between a circular enclosure, formed by uprights on the deck behind the capstan. Ropemakers and artificers examined the rope which had been already used. They served the injured strands with yarn, renewed portions chafed to death, tested bolts and shackles and swivels, and bent on new lengths of rope and hawser, whilst the ship was proceeding to take up her position for another demonstration against the Cable. The line now employed, the last left in the ship, was a thing of shreds and patches. It consisted of 1,600 fathoms of wire rope, 220 fathoms of hemp, and 510 fathoms of Manilla hawser, of which 1,760 fathoms could be depended upon, the rest being "suspicious." The morning was not very fine; but the wind was light, and on the whole favourable, and the only circumstance to cause doubt or uneasiness was the current, the influence of which could not be determined. The observations of the officers rendered it doubtful whether the buoy No. 2 had drifted, and it was rather believed that in the interval between the breaking of the grapnel and the letting-go of the buoy, the Great Eastern herself had drifted from the place, and thus caused the apparent discrepancy in position. At 7·45 a.m. the ship was alongside buoy No. 2 once more, and thence proceeded

E. Walker. lith. from a drawing by R. Dudley

London. Day & Son, Limited. Lith.

FORWARD DECK CLEARED FOR THE FINAL ATTEMPT AT GRAPPLING _ AUGUST 11TH

to an advantageous bearing for drifting down on the Cable with her grapnel. The Terrible kept about two miles away, regarding our operations with a melancholy interest. At 11·30 a.m., ship's time, the Great Eastern signalled "We are going to make a final effort," and soon afterwards, "We are sorry you have had such uncomfortable waiting." At 1·56 p.m., Greenwich time, when buoy No. 2 was bearing E. by N. about two miles, the ship's head being W. and by S., the grapnel was let go, and soon reached the bottom, as the improvements in the machinery and capstan enabled the men to pay it out at the rate of fifty fathoms a minute. The fore-and-aft canvas was set, to counteract the force of the current, and the Great Eastern drifted to N.E., right across the Cable, before a light breeze from S.W. At first there was only a strain of 42 cwt. shown, and the ship went quite steadily and slowly towards the Cable. At 3·30 p.m. the strain increased, and then the Great Eastern gave some little sign of feeling a restraint on her actions from below, her head describing unsteady lines from W.N.W. to W. by S. The screw engines were gently brought into play to keep her head to the wind. The machinery and capstan, which had been put in motion some time previously to haul in the grapnel cable, now took it in easily and regularly, except when a shackle or swivel jarred it for a moment. Every movement of the ship was most keenly watched, till the increasing strain on the dynamometer showed that the same grip on the bottom which had twice turned the head of the Great Eastern, was again placed on the grapnel she was dragging along the bottom of the Atlantic. The index of the dynamometer rose: it marked 60 cwt., then it jerked up to 65 cwt., then it reached 70 cwt., then 75 cwt.: at last its iron finger pointed to 80 cwt. It was too much to stand by and witness the terrible struggle between the crisping, yielding hawser, which was coming in fast, the relentless iron-clad capstan, and the fierce resolute power in the black sea, which seemed endued with demoniacal energy as it tugged and swerved to and fro on the iron hook. But it was beyond peradventure that the Atlantic Cable had been hooked and struck, and was coming up from its oozy bed. What alternations of hope and fear—what doubts, what sanguine dreams, dispelled by a moment's thought, only to revive again! What need to say how men were agitated on board the ship? There was in their breasts, those who felt at all, that intense quiet excitement with which we all attend the utterance of a supreme decree, final and irrevocable. Some remained below in the saloons—fastened their eyes on unread pages of books, or gave expression to their feelings in fitful notes from piano or violin. Others went aft to the great Sahara of deck where all was lifeless now, and whence the iron oasis had vanished. Some walked to and fro in the saloon; others paced the deck amidships.   None liked to go forward, where every jar of the machinery, every shackle that passed the drum, every clank, made their hearts leap into their mouths.   Captain Anderson, Mr. Canning, Mr. Clifford, and the officers and men engaged in working the ship and taking in the grapnel, were in the bows of course, and shared in the common anxiety.   At dinner-time 500 fathoms of grapnel rope had been taken in, and the strain was mounting beyond 82 cwt. Nothing else could be talked of.   The boldest ventured to utter the words "Heart's Content" and "Newfoundland" once more.   All through the unquiet meal we could hear the shrill whistle through the acoustic tube from the bow to the bridge, which warned the quartermasters to stop, reverse, or turn ahead the screw engines to meet the exigencies of the strain on the grapnel rope. The evening was darkling and raw.   At 6·30 I left the saloon, and walked up and down the deck, under the shelter of the paddle-box, glancing forward now and then to the bow, to look at the busy crowd of engineers, sailors, and cablemen gathered round the rope coming in over the drum, which just rose clear of one of the foremasts, and listening to the warning shouts as the shackles came inboard, and hurtled through the machinery till they floundered on the hurricane deck.

About 20 minutes had elapsed when I heard the whistle sound on the bridge, and at the same time saw one of the men running aft anxiously. "There's a heavy strain on now, sir," he said.   I was going forward, when the whistle blew again, and I heard cries of "Stop it!" or "Stop her!" in the bows, shouts of "Look out!" and agitated exclamations.   Then there was silence.   I knew at once all was over. The machinery stood still in the bows, and for a moment every man was fixed, as if turned to stone . there, standing blank and mute, were the hardy constant toilers, whose toil was ended at last.   Our last bolt was sped.   Just at the moment the fracture took place, Staff-Commander Moriarty had come up from his cabin to announce that he was quite certain, from his calculations, that the vessel had dragged over the Cable in a most favourable spot.   It was 9·40 p.m., Greenwich time, and 765 fathoms had been got in, leaving little more of the hempen tackle to be recovered, when a shackle came in and passed through the machinery, and at the instant the hawser snapped as it was drawn to the capstan, and, whistling through the air like a round shot, would have carried death in its course through the crowded groups on the bows, but for the determination with which the men at the stoppers held on to them, and kept the murderous end straight in its career, as it sped back to the Atlantic.   It was scarcely to be hoped that it had passed harmlessly away.   Mr. Canning and others rushed forward, exclaiming, "Is any one hurt?" ere the shout "It is gone!" had subsided. The battle was over! Then the first thought was for the wounded and the dead, and God be thanked for it, there were neither to add to the grief of defeat. Nigh two miles more of iron coils, and wire, and rope were added to the entanglement of the great labyrinth made by the Great Eastern in the bed of the ocean. In a few seconds every man knew the worst. The bow was deserted, and all came aft and set about their duties. Mr. Clifford, with the end of a hempen hawser in his hand, torn in twain as though it were a roll of brown paper —Mr. Canning already recovered from the shock, and giving orders to stow away what had come up from the sea—Captain Anderson directing the chief engineer to get up steam, and prepare for an immediate start.

The result was signalled to the Terrible, which came down to us, and as she was bound to St. John's to take in coals to enable her to return to England, all who had business or friends in America prepared their dispatches for her boat. The wind and sea were rising, as if anxious to hurry us from the scene of the nine days' struggle. The Great Eastern's head was already turned westwards. All were prompt to leave the spot which soon would bear no mark of the night and day long labours—for the buoys which whirled up and down and round in the seaway would probably become waifs and strays on the ocean, and all that was left of the expedition for a time were the entries in log books—" Lat. 51 24' Long. 38° 59'; end of Cable down N. 50 W. $1\frac{3}{4}$ mile "—and such memories as animate men who, having witnessed brave fights with adverse fortune, are encouraged thereby to persevere, in the sure conviction that the good work will in the end be accomplished. It was wild and dark when Lieutenant Prowse set off to regain his ship. The flash of a gun from the Terrible to recall her cutter lighted up the gloom, and the glare of an answering blue light, burned by the boat, revealed for an instant the hull of the man-of-war on the heaving waters. There was a profound silence on board the Big Ship. She struggled against the helm for a moment as though she still yearned to pursue her course to the west, then bowed her head to the angry sea in admission of defeat, and moved slowly to meet the rising sun. The signal lanterns flashed from the Terrible, " Farewell!" The lights from our paddle-box pierced the night, " Good-by! Thank you," in sad acknowledgment. Then each sped on her way in solitude and darkness.

The progress of the undertaking excited the utmost interest, not only in Great Britain, but over all the civilised world. Twice a day the telegraph at Foilhummerum spread to all parts of the earth a brief account of the doings of the Great Ship. Almost as soon as one of the unexpected impediments which marred the successful issue of the enterprise arose, the public were informed of it, and could mark on the map the spot where sailor, engineer, and electrician were engaged in their work on the bosom of the wide Atlantic ere their labours were over. The Great Eastern's position could be traced on the chart, and the course of the Cable, in its unseen resting-place, could be followed from day to day. The "faults" caused more surprise perhaps on shore than on board, because those engaged in paying-out the Cable were re-assured by the certainty with which the faults were detected, and the comparative facility with which the Cable was taken up from the sea. Although the various delays which occurred produced some discouragement and uneasiness among those who had worked so hard and embarked so much in the grand project, the ease with which communication was restored as often as it was injured or interrupted by faults and dead earth, inspired confidence in the eventual success of the attempt. But only those actually witnesses of the wonderful facility with which the Cable was paid out felt the conviction that the Cable could be laid. The public only knew the general results, and did not appreciate properly the nature of the difficulties to which the frustration of their hopes was due. When the last fault occurred, the electricians at Valentia were left without any precise indications of the nature of the obstruction, or of the proceedings of those on board; but they actually calculated within a few fathoms the exact locality of the injury; and when the end of the Cable sank into the depths of the ocean, the practical wizards of Foilhummerum could tell where it was to be found, though they could not see and could not hear. When all communication ceased with the Great Eastern no uneasiness was excited, because a similar event had occurred before for many hours, and the ship spoke after all. But hour after hour passed away on leaden wings, and day followed day, and the needle was still, and the light moved not in the darkened chamber at Foilhummerum. It may be conceived with what solicitude the men, in whose watchfulness all the sleeping and waking world were interested, looked out for some sign of the revival of the current in the dull veins of the subtle mechanism.

The directors and shareholders of the two companies represented something more than the enormous stake they had put in the undertaking. Their feelings were shared by the mass of the people, and Her Majesty was animated by the same solicitude as her subjects. For there had been prophets of evil before the expedition sailed, and now their voices were raised again, and found credence among those who distrusted the magnificent ship which was then calmly breasting the billows of the Atlantic—the envy of her guardians—as well as among the class whose normal condition is despair of every scheme, good, useful, novel, or great. The newspapers began to admit speculations and argumentative letters into their columns, and although the original articles did not indicate any apprehension of a catastrophe, it was evident the public mind was becoming uneasy. The feeling increased. The correspondence augmented in volume, and, let it be said, in wildness of conjecture and unsoundness of premises and conclusions. Those who were inclined to believe that the Great Eastern had gone to the bottom were comforted by the reflection that the two men-of-war would save those who were on board. Had they known that the Sphinx had disappeared, and that the Great Eastern was much better able to help the Terrible, in a time of watery trouble, than the Terrible would be to aid her, they would have despaired indeed.

All the while those on board engaged in their work—grappling and lifting, drifting and sailing—were enjoying themselves as far as the uncertainty attendant on their work would allow them, and were in a state of repose barely disturbed, as the time wore on, by surmises that people at home might begin to entertain doubts as to what had become of the expedition. Even these speculations would have had no agitating influence had the electricians on board communicated with the shore before they cut the end of the Cable on the last occasion. It would have surprised and amused officers and crew if they could have known that the vessel, which they were never tired of praising and admiring, was pronounced by eminent engineers to need strengthening ; that she had sunk in the middle, or had fagged ; or if they could have read confident assertions that the grand fabric in which they were so comfortably lodged and entertained and borne was unsafe and radically faulty ; that good authorities had declared she was hogged. Undoubtedly there were grounds of anxiety, but none for anticipations and predictions of the worst. It would not be fair to omit to mention that in some instances the most correct and close conjectures were made concerning the position of the ship and the work in which she was engaged, as well as the causes of the long-continued silence. Several letters appeared, in which the writers tried, with singular justice of reasoning, to stem the current of alarm. The press generally abstained from any adverse speculations ; but it was rather behind the public feeling in that respect. It cannot be denied that the news-agent who hailed the Great Eastern at Crookhaven with the words, " We did not know what to make of you. Many think you went down," expressed the conviction of a great number of persons all over the kingdom, on the 17th August.

Early on the morning of that day the Great Eastern came in sight of land, and soon after 7 o'clock a.m. steamed into Crookhaven, to land a few passengers and to communicate with the telegraph station at that solitary and romantic spot. Ere noon the news of the safety of the ship relieved many an anxious thought, silenced many a tongue and pen, and dissipated many a gloomy apprehension. It may be said that the return of the Great Eastern was a subject of national rejoicing. Every newspaper in the kingdom contained articles on the topic. The narrative of the voyage, which was written on board, and sent to all the principal journals before the Great Eastern arrived at the Nore, so that the public were at once placed in possession of every fact connected with the proceedings, almost simultaneously, was read with the utmost avidity, and when the facts were known, all men concurred in the justice of the leading articles which, without exception of note, drew fresh hopes of success from the record of the causes which led to the interruption of the enterprise. The energy, skill, and resolution displayed in the attempt to recover the Cable were admitted and praised on all hands. But what most excited attention was the fact that the Cable had actually been hooked three times at a depth of two nautical miles, and carried up halfway to the top. The most sceptical were convinced when they became aware of the hard material evidence on that point. Next in point of interest perhaps was the conduct of the Great Eastern herself. A great revulsion of sentiment took place in favour of the vessel which had hitherto been unfortunate in her management, or in the conditions under which she had been tried.

Whilst the most profound ignorance respecting the fate of the Great Eastern prevailed, an Extraordinary General Meeting of the Atlantic Telegraph Company was held on 8th August, in pursuance of a notice issued on 24th July previous, to consider the expediency of converting into Consolidated Eight per Cent. Preferential Stock the Eight per Cent. Preferential Capital of the Atlantic Telegraph Company, consisting of 120,000 shares of 5l. each, and of converting into Ordinary Consolidated Stock the whole of the Ordinary Share Capital, consisting of 350 shares of the par value of 1000l., and 5,463 shares of the par value of 20l., and to issue either in ordinary stock or in shares the sum of 137,140l. of ordinary capital, authorised at the Extraordinary General Meeting of March 31st, 1864, and agreed to be issued in instalments fully paid up, to the contractors from time to time after the successful completion of their contract.

The directors also gave notice that they intended to seek authority from the shareholders to issue such amounts of new capital as may be required for the construction and laying of a second Atlantic Telegraph Cable under powers of their Act of Parliament, and to attach to such capital such privileges and such advantages and conditions as might be determined. The Right Hon. J. S. Wortley, chairman, who has exhibited unshaken confidence and untiring energy in the post he occupies, had a difficult task before him, but even then he could exhort his hearers to courage and perseverance. As he well said, " But there are two things from which we may derive considerable consolation. This great enterprise has been the subject of discussion in every civilised nation in the world. The eyes of science have been fixed upon it; and the acuteness of criticism has been brought to bear on it. We have had our detractors, and there have been sceptics; and what are the two main points on which they have founded their scepticism? One is, that the great depth of nearly three miles must bring extraordinary pressure on the Cable, must injure it by perforating the covering, and must in fact destroy the insulation. The other point was the impossibility, as they contended, of communicating intelligible signals through so great a length, or 'leap' as they term it, as 1,600 miles. But we had a scientific committee, who made experiments, and who assured themselves that there was nothing in either of those objections; and now we have in addition the much more practical and valuable proof of experience. What are the facts? Some days before the interruption of the messages the Great Eastern passed over the deepest portion of the ocean (with one slight exception) which we have to traverse between Europe and America. She passed safely over a depth of 2,400 fathoms, telegraphing perfect signals. This entirely disproves and refutes the first objection and doubt which existed in the minds of those sceptical gentlemen, because the Cable was laid in great depths, varying from 1,500 to 2000 fathoms, and even in 2,400 fathoms; and so far from the great pressure at that depth injuring the Cable, the Company's signals appear from their telegrams to have improved every yard they went; and the signals through 2,400 fathoms of water were as perfect as, if not more perfect than, those at a less depth. That is in confirmation of the old Cable having worked at those depths. Then I say that our scientific committee, and those who said that the pressure would not have an injurious effect, have been fully borne out; and that the result has proved that, so far from injuring it, pressure improves the Cable. In spite of these facts, I see here a communication from a gentleman to one of the public journals only yesterday, in which he says, that looking at the pressure of a column of water equal to so many atmospheres, it must destroy the Cable; and he adds with confidence, that the Cable must be at the present moment a perfect wreck! And then he says that the Company never made experiments to satisfy themselves what this number of atmospheres would do to the Cable. He writes in perfect ignorance, that the scientific committee has the means afforded them by this Company of applying a weight of 6000lb. to the square inch; but after having proceeded to a certain extent with that experiment, and tried a very large amount of pressure, and finding that the Cable, so far from deteriorating, was improved by the compression of its elements, they thought it unnecessary to carry the experiments further. And now we have the result to corroborate their views."

On October 12, an Extraordinary General Meeting of the Atlantic Telegraph Company was held, at which the Chairman, the Right Hon. J. S. Wortley, proposed a Resolution rescinding those passed at the General Meeting in August. He reminded them the Capital was originally issued in 1000*l*. shares. After that an additional amount of capital was raised in 20*l*. shares ; and after the first failure a further capital of 600,000*l*. in 5*l*. shares, and an 8 per cent. preference, was raised. Under these circumstances they succeeded in raising the necessary sum enabling them to send out the last expedition, and they now proposed that notwithstanding that guarantee of 8 per cent. to issue a new preferential capital at the rate of 12 per cent. They had negotiated with the same contractors who had hitherto had charge of laying the Cable, and they were willing for the sum of 500,000*l*. to take out a sufficient quantity of Cable, together with that which was left in the ship amounting to about 1000 miles, and in the first place to go across and lay a new Cable, and then to come back and pick up the old one, splice it, and continue it to Newfoundland. He might say at once, that not only the contractors, but all who were engaged in the undertaking, were represented there that day, as well as the able staff of scientific men to whom they were so much indebted upon the last expedition, and he said in their presence that they all had extreme confidence that they would not only be able to lay the new Cable but to pick up the old one, mend it, and relay it. It was proposed that in addition to the 500,000*l*. there should, if the Cable was successfully laid, be a contingent profit to the contractor, which would be paid in money. It was apprehended that the additional 100,000*l*. asked for would be quite sufficient to meet any contingency that might arise. The formal Resolutions rescinding those passed at the meeting in August last were carried unanimously ; and it was Resolved, " That the Capital of the Company be increased to an amount not exceeding 2,000,000*l*., by the creation and issue of not exceeding 160,000 new shares of 5*l*. each, and that such new shares shall bear and be entitled to a preferential dividend at the rate of 12*l*. per cent. per annum on the amount for the time being paid up thereon, in priority to any dividend or on any other capital of the Company, and shall also, in proportion to the amount for the time being paid up thereon, be entitled to participate equally with the other capital of the Company in any moneys applicable to dividend, which upon each declaration of dividend may remain after paying or providing for the said dividend of 12*l*. per cent. per annum, the preferential dividend of 8*l*. per cent. per annum payable on the consolidated 8 per cent. preferential stock of the Company, and a dividend at the rate of 4 per cent. per annum on the consolidated ordinary stock and ordinary shares of the Company."

In their Prospectus, the Directors stated that the Telegraph Construction and Maintenance Company, in consideration of the sum of 500,000*l.*, which has been agreed on as the cost price of the Cable if paid for in cash, have already commenced the manufacture of the new Cable, to be laid down during 1866 between Ireland and Newfoundland. The contractors, if the said Cable be successful, but not otherwise, are to have in shares and cash a profit at the rate of 20 per cent. upon the cost. The contractors also undertake during 1866, without any further charge whatever, to go to sea with sufficient Cable, including that now left on board the Great Eastern, and all proper appliances and apparatus such as experience has shown to be necessary, and to use their best endeavours—in the success of which they express entire belief—to recover, repair, and complete in working order between Ireland and Newfoundland, the present broken Cable, which has been ascertained by recent careful electrical tests to be in perfect order throughout its entire length. It will be seen that circumstances have thus enabled the Board to effect a very considerable economy in the Company's present operations, for in the event of success the Company will be in possession of two efficient Cables for a considerably less amount than would have been expended if the Cable of this year had been successfully laid, and another had been purchased separately. Subscriptions were invited for the sum of 600,000*l.*, in 120,000 shares of 5*l.* each.

This new capital will not only create fresh property, but probably resuscitate the old; and the experience of the present year shows that by these means the existing 8 per cent. Preference Stock will, in all probability, be again placed at par in the market before the sailing of the ship next year.

These new Shares will accordingly be entitled to take precedence as to dividend over all the other existing stock of the Company, and to participate *pro ratá* in all subsequent dividends, bonuses, or benefits, after 8 per cent. shall have been paid upon the second preference stock and 4 per cent. upon the ordinary stock.

The profits to be expected on the completion of this work, if each of the two proposed Cables can be worked at the very low rate of only five words per minute upon each Cable for sixteen hours a day at five shillings per word, the traffic, after paying the dividend charges of 12, 8, and 4 per cent. respectively, amounting together to 144,000*l.* upon the capital comprised in those different stocks, and after paying the very large sum of 50,000*l.* a year for working expenses, would leave a very large balance for paying further dividends or bonuses on the Company's total capital, both ordinary and preferential, or for reserve funds if preferred.

A calm examination of the courses which led to the suspension of the Great

Eastern's work, inspired those whose judgments were free from prejudice with the belief that a series of accidents, in their nature easily guarded against in future, had been the sole causes of the frustration of the enterprise. If the external coating had not been injured, no faults could have occurred, and if there had been no faults, the Cable would have been laid with the utmost ease. The success of the Telegraph becomes assured the moment the occurrence of faults can be obviated, or their detection can be followed by immediate reparation. These objects are to be attained, and the Directors, encouraged by the confidence of the public, and by the enormous gains which must reward even a temporary success, set about to secure them. An arrangement was entered into with the Directors of the Great Ship Company by which the Telegraph Construction and Maintenance Company secured the Great Eastern for a term of years, and another negotiation ended in obtaining the services of Captain Anderson in charge of her.

Now it may be fairly concluded, from our experience of the " Atlantic Telegraph Expeditions " in 1857, 1858, and 1865,—That a submarine telegraph Cable can be laid between Ireland and Newfoundland, because it was actually done in 1858. That messages can be transmitted through a Cable so laid, because 271 messages were sent from Newfoundland to Valentia, and 129 messages from Valentia to Newfoundland, in 1858. That the insulation of a Cable increases very much after its submersion in the cold deep water of the Atlantic, and that its conducting power is considerably improved thereby. That the steamship Great Eastern, from her size and constant steadiness, and from the control over her afforded by the joint use of paddle and screw, renders it possible and safe for her to lay an Atlantic Cable without regard to the weather. That the egress of a Cable in the course of being laid from the Great Eastern may be safely stopped on the appearance of a fault, and with strong tackle and good hauling-in machinery, the fault may be lifted from a depth of over 2000 fathoms, and cut out on board the ship, and the Cable respliced and laid in perfect condition. That in a depth of two miles a Cable can be caught at the bottom, because four attempts were made to grapple the Cable in 1865, and in three of them the Cable was caught by the grapnel.

The paying-out machinery, constructed by Messrs. Canning and Clifford, and used on board the Great Eastern in 1865, worked perfectly, and can be confidently relied on for laying Cables across the Atlantic. With the improved telegraphic instruments, for long submarine lines, of Professor W. Thomson and Mr. Varley, a speed of more than eight words per minute can be obtained through such a circuit as the Atlantic Cable of 1865, between Ireland and Newfoundland ; as the amount of slack actually payed-out did not exceed 14 per cent., which would have made the total Cable laid between Valentia and Heart's Content less than 1,900 miles.

The Cable of 1865, though capable of bearing a strain of 7 tons, did not experience more than 14 cwt. in being payed-out into the deepest water of the Atlantic between Ireland and Newfoundland.

There is no difficulty in mooring buoys in the deep water of the Atlantic between Ireland and Newfoundland ; a buoy, even when moored by a piece of the Atlantic Cable itself which had been previously lifted from a depth of over 2000 fathoms, has ridden out a gale.

More than four miles of the Atlantic Cable have been recovered from a depth of over two miles, and the insulation of the gutta-percha-covered wire was in no way whatever impaired, either by the depth of water or the strains to which it had been subjected by lifting and passing through the hauling-in apparatus.

The Cable of 1865, owing to the improvements introduced into the manufacture of the gutta percha, insulated more than one hundred times better than Cables made in 1858, then considered perfect, and still working. The improvements effected since the beginning of 1851 in the conducting power of the copper wire, by selecting it, has increased the rate of signalling possible through long submarine Cables by more than 33 per cent. Electrical testing can be conducted at sea with such certainty as to discover the existence of faults in less than a minute of their occurrence. If a steam-engine be attached to the paying-out machinery, so as to permit of hauling-in the Cable immediately a fault is discovered, and a slight modification made in the construction of the external sheath of the Cable, the cause of the faults experienced will be entirely done away with ; and should a fault occur, it can be picked up even before it has reached the bottom of the Atlantic.

The Great Eastern is now undergoing the alterations which will render her absolutely perfect for the purpose of laying the new Cable and picking up the old, and next year will see the renewal of the enterprise of connecting the Old World with the New by an enduring link which, under God's blessing, may confer unnumbered blessings on the nations which the ocean has so long divided, and add to the greatness and the power which this empire has achieved by the energy, enterprise, and perseverance of our countrymen, directed by Providence, to the promotion of the welfare and happiness of mankind. Remembering all that has occurred,—how well-grounded hopes were deceived, just expectations frustrated, —there are still grounds for confidence, absolute as far as the nature of human affairs permits them in any calculation of future events to be, that the year 1866 will witness the consummation of the greatest work of civilised man, and the grandest exposition of the development of the faculties bestowed on him to overcome material difficulties.

The last word transmitted through the old Telegraph from Europe to America, was " Forward," and " Forward " is the motto of the enterprise still.

FINIS.

APPENDIX.

APPENDIX.

———◆———

A.

The following is a list of the Gentlemen connected with the project for the year 1865

————

NEW YORK, NEWFOUNDLAND, AND LONDON TELEGRAPH COMPANY.

PETER COOPER, Esq. . . . . . PRESIDENT.
CYRUS W. FIELD, Esq. . . . . . VICE-PRESIDENT.
MOSES TAYLOR, Esq. . . . . . TREASURER.
PROF. S. F. B. MORSE , ... ELECTRICIAN.
DAVID DUDLEY FIELD, Esq. . . . . COUNSEL.

DIRECTORS.

PETER COOPER, Esq. . . ⎫
MOSES TAYLOR, Esq . . . ⎪
CYRUS W. FIELD, Esq. . . ⎬ NEW YORK.
MARSHALL O. ROBERTS, Esq.. ⎪
WILSON G. HUNT, Esq. . . ⎭

SECRETARY.

ROBERT W. LOWBER, Esq.

GENERAL SUPERINTENDENT.

ALEXANDER M. MACKAY, Esq., St. John's, Newfoundland.

~~~~~~~~~~~~~~~~~~~~~~~

ATLANTIC TELEGRAPH COMPANY.

DIRECTORS.

THE RIGHT HON. JAMES STUART WORTLEY, *Chairman.* | CURTIS M. LAMPSON, Esq., *Vice-Chairman.*

G. P. BIDDER, Esq., C.E.	SIR EDWARD CUNARD, Bart.	EDWARD MOON, Esq.
FRANCIS LE BRETON, Esq.	SAMUEL GURNEY, Esq., M.P.	GEORGE PEABODY, Esq.
EDWARD CROPPER, Esq.	CAPTAIN A. T. HAMILTON.	JOHN PENDER, Esq., M.P.

HONORARY DIRECTOR—W. H. STEPHENSON, Esq.

HONORARY DIRECTORS IN THE UNITED STATES.

E. M. ARCHIBALD, Esq., C.B., H.M. Consul, New York.	CYRUS W. FIELD, Esq. . . . . . New York.
PETER COOPER, Esq. . . . . . New York.	WILSON G. HUNT, Esq. . . . . New York.
WILLIAM E. DODGE, Esq. . . . . New York.	A. A. LOW, Esq. . . . . . . New York.

HOWARD POTTER, Esq., New York.

HONORARY DIRECTORS IN BRITISH NORTH AMERICA.

HUGH ALLEN, Esq., Montreal, Canada.	WALTER GRIEVE, Esq., St. John's, Newfoundland.
WILLIAM CUNARD, Esq , Halifax, Nova Scotia.	THOMAS C. KINNEAR, Esq., Halifax, Nova Scotia.

APPENDIX.

ATLANTIC TELEGRAPH COMPANY—*continued.*

CONSULTING SCIENTIFIC COMMITTEE.

WILLIAM FAIRBAIRN, Esq., F.R.S., Manchester.   PROFESSOR WM. THOMSON, F.R.S., Glasgow.

CAPTAIN DOUGLAS GALTON, R.E., F.R.S., London.   PROFESSOR C. WHEATSTONE, F.R.S., London.

JOSEPH WHITWORTH, Esq., F.R.S., Manchester.

HONORARY CONSULTING ENGINEER IN AMERICA—GENERAL MARSHALL LEFFERTS, New York.

Offices—12, *St. Helen's Place, Bishopsgate Street Within, London.*

SECRETARY AND GENERAL SUPERINTENDENT—GEORGE SAWARD, Esq.

ELECTRICIAN—CROMWELL F. VARLEY, Esq.   SOLICITORS—MESSRS. FRESHFIELDS & NEWMAN.

AUDITOR—H. W. BLACKBURN, Esq., Bradford, Yorkshire, Public Accountant.

BANKERS.

In London—The Bank of England, and Messrs. Glyn, Mills, & Co.

In Lancashire—The Consolidated Bank, Manchester.

In Ireland—The National Bank and its Branches.

In Scotland—The British Linen Company and its Branches.

In New York—Messrs. Duncan, Sherman, & Co.

In Canada and Nova Scotia—The Bank of British North America.

In Newfoundland—The Union Bank of Newfoundland.

B.

THE TELEGRAPH CONSTRUCTION AND MAINTENANCE COMPANY

(Uniting the Business of the Gutta Percha Company with that of Messrs. Glass, Elliot, & Company)

is constituted as follows :—

DIRECTORS.

JOHN PENDER, Esq., M.P., *Chairman.*

ALEXANDER HENRY CAMPBELL, Esq., M.P., *Vice-Chairman.*

RICHARD ATWOOD GLASS, Esq. (Glass, Elliot, & Co.), *Managing Director.*

HENRY FORD BARCLAY, Esq. (Gutta Percha Co.)	DANIEL GOOCH, Esq., C.E., M.P.
THOMAS BRASSEY, Esq.	SAMUEL GURNEY, Esq., M.P.
GEORGE ELLIOT, Esq. (Glass, Elliot, & Co.)	LORD JOHN HAY.
ALEXANDER STRUTHERS FINLAY, Esq., M.P.	JOHN SMITH, Esq. (Smith, Fleming, & Co.)

BANKERS—THE CONSOLIDATED BANK, London and Manchester.

SOLICITORS.

MESSRS. BIRCHAM, DALRYMPLE, DRAKE, & WARD. | MESSRS. BAXTER, ROSE, NORTON, & Co.

SECRETARY—WILLIAM SHUTER, Esq.

Offices—54, *Old Broad Street, London.* *Works—Wharf Road, City Road, N., and East Greenwich, S.E.*

C.

THE following will be some of the Improvements in the Picking-up Machinery
and in the Vessel to fit her for her next voyage, and it is believed that the
Great Eastern will be as perfect and as admirably adapted for her work
as human hands can make her.

The whole apparatus will be strengthened and improved by grooved drums, and more boiler
power added, and other drums will be provided for lowering away buoy-rope when grappling.

The paying-out machinery will have steam-power added to it, the spare drum fitted on the
machine will be used for picking-up in connection with the paying-out drum ; an extra drum and
brake-wheel will also be placed near the stern for the purpose of paying-out grapnel lines and
buoy-rope, in case it is found more convenient than at the bow.

The grapnel-rope, with shackles, swivels, &c., will be made sufficiently strong to lift or break
the bight of the Cable in the deepest water.   The hawse-pipes and stem of the ship will be guarded
to prevent the Cable from being injured.   A guard will be placed round the screw to prevent the
Cable and buoy-rope fouling.

D.

STATEMENT OF KNOTS RUN AND CABLE PAYED-OUT PER DAY.

Sunday, July 23.—Left Berehaven at 1·45 a.m. Passed Skelligs at 8·0 a.m. ; bore away N.W., and came up with Caroline at 8·30 a.m., about 25 miles N.W. of Valencia. 10·30 a.m., End got out of afterhold. 11·0 a.m., Terrible and Sphinx came alongside. 12·35 p.m., Caroline got up end of shore-end Cable. 12·45 p.m., passed end of deep-sea Cable to Caroline over stern-sheave of Great Eastern. 5·20 p.m., splice finished on board Caroline, and bight of Cable slipped. 6·50 p.m., took hands on board from Caroline. 8·0 p.m., paddle and screw engines started.

Date. 12 Noon.	Made Good. Course.	Dist.	Lat. N. Obs.	Long. W. Obs.	Distance from Valencia	Miles payed-out.	Slack per Cent.	Heart's Content. Bearing.	Distance.
July 23	Splice to Shore end		51 50 0	11 2 20	24½	27·00	—	N. 80., W.	1638·5
24	{ Picking up Cable }		52 2 30	12 17 30	73·1	84·791	15.99	—	—
25			51 58 0	12 11 0	68·5	74·591	8·89	—	1596·5
26	N. 79., 20. W.	111·5	52 18 42	15 10 0	180	191·96	6·64	N. 24., 21 W.	1485
27	N. 81., 30. W.	142·5	52 34 30	19 0 30	320·8	357·55	11·45	N. 87., 39 W.	1344·2
28	N. 86., 30. W.	155·5	52 45 0	23 15 45	476·4	531·57	11·16	S. 88., 35 W.	1188·6
29	S. 87., 40. W.	160·0	52 38 30	27 40 0	636·4	707·36	11·15	S. 84., 54 W.	1028·6
30	S. 70., 0. W.	24	52 30 30	28 17 0	659·6	745·0	12·94	S. 84., 48 W.	1005·4
31	S. 81., 0. W.	134	52 9 20	31 53 0	793	903·0	15·13	S. 82., 20 W.	871·9
Aug. 1	S. 83., 45. W.	155	51 52 30	36 3 30	948	1081·55	14·09	S. 78., 22 W.	717·1
2	{ S. 76., 25 W. 115·4 } Returned 2 miles before Cable broke		51 25 0	39 1 0	1063·4	1186·0	11·56	S. 76., 17 W.	603·6
3	—	—	DR. 51 36 0	38 27 0	—	—	—	—	—
4	—	—	OBS. 51 34 30	37 54 0	—	—	—	End of Cable.	S.76.,W.,44 M.
5	—	—	51 25 0	38 36 0	—	—	—	,, ,,	W. (true) 15 M.
6	—	—	OBS. 51 25 0	38 20 0	—	—	—	,, ,,	W. ,, 26 M.
7	—	—	51 29 30	39 4 30	—	—	—	,, ,,	S. 23., E., 5 M.
8	—	—	51 28 0	38 56 0	—	—	—	No. 2 Buoy	W.S.W., 3 M.
9	—	—	51 29 30	39 6 0	—	—	—	,, ,,	S. 38, 6 or 7 M.
10	—	—	51 26 0	38 59 0	—	—	—	End of Cable	S. 56., W., 2 M.
11	—	—	51 24 0	38 59 0	D.R.	—	—	,, ,,	N. 50, W. 1¾ M.

TEMPERATURE OF THE SEA-WATER.

Date.	Time.	Degrees.	Date.	Time.	Degrees.
1865. July 26th	Noon.	59	1865. August 4th	Noon.	55
,, 27th	,,	65	,, 5th	,,	55
,, 28th	,,	56	,, 6th	,,	55
,, 29th	,,	55	,, 7th	,,	54
,, 30th	,,	53	,, 8th	,,	55
,, 31st	,,	56	,, 9th	,,	54
August 1st	,,	59	,, 10th	,,	57
,, 2nd	,,	59	,, 11th	,,	57
,, 3rd	,,	54	,, 12th	,,	54

S. CANNNNG.

APPENDIX.

E.

THE FOLLOWING IS A TABLE OF THE CABLES ALREADY LAID IN THE SEAS AND OCEANS OF THE WORLD.

No.	Cable.	Iron.		lbs.	Copper.		Length of Cable.
		Weight.	Length.	G. P.	lbs.	Length.	
1	Dover and Cape Grisnez			13,230	3300	30	30
2	Dover and Calais . .	314,600	260	14,820	7060	104	26
3	Holyhead, Howth . .	156,480	960	11,400	5400	80	80
4	Portpatrick and Donaghadee	316,200	300	20,312	10,125	150	25
5	Denmark	164,748	162	5400	2052	54	18
6	Dover, Ostend	1,138,320	1080	73,125	36,450	540	90
7	Frith of Forth . . .	77,800	200	8180	18,520	20	5
8	Italy, Corsica . . .	1,597,200	1320	104,940	44,550	660	110
9	Corsica, Sardinia . . .	145,200	120	9540	4050	60	10
10	Holyhead, Howth . .	295,640	760	15,504	51,300	76	76
11	Do.	295,640	760	15,504	51,300	76	76
12	Portpatrick and Whitehead	328	848	312	22,280	10,530	16s 284
13	Sweden, Denmark . .	137,020	130	5558	2633	39	13
14	Black Sea			56,763	24,098	357	357
15	Do.	70,584	2076	24,652	11,678	173	173
16	Prince Edward's Island, New Brunswick	46,512	144	1905	1134	84	12
17	England, Hanover . .	807,680	3360	66,360	30,240	2240	280
18	— Holland . . .	2,439,840	1366	110,976	78,336	544	136
19	Liverpool, Holyhead .	161,400	300	5925	3376	50	25
20	Channel Islands . . .	450,306	837	14,787	10,230	93	93
21	Isle of Man	193,680	360	7344	2430	36	36
22	England, Denmark . .	2,734,200	4200	124,425	6700	4200	350
23	Folkestone, Boulogne .	429,120	288	20,520	7776	576	24
24	Singapore, Batavia . .	564,300	9900	112,200	86,350	3850	550
25	Sweden, Gottland . .	248,064	768	10,176	6048	448	64
26	Tasmania	933,600	2400	38,160	16,480	240	240
27	Denmark, Great Belt .	203,280	168	13,365	5628	84	14
28	Dacca, Pegu	119,016	2088	21,228	18,096	812	116
29	Newfoundland, Cape Breton	290,700	900	13,515	8500	595	85
30	First Atlantic	5,140,800	428,400	748,000	340,000	23,800	3400
31	Sardinia and Malta: Dardanelles to Scio and Candia from Scio,	3,326,400	12,600	111,300	70,000	4900	700
32	Athens, to Syra and Scio	631,104	8304	82,521	51,900	3633	519
33	Sardinia, Bona . . .	707,000	1500	42,750	80,000	500	125
34	Red Sea and India . .	6,126,714	63,168	743,908	547,404	24,563	3509
35	Sicily and Malta . . .	499,100	700	10,080	7000	490	70
36	Barcelona, Mahon . .	538,560	2880	25,920	16,740	1260	180
37	Iviza to Majorca: St. Antonia to Iviza .	639,900	2700	31,800	18,000	1200	150
38	Toulon, Algiers . . .	465,600	4800	93,600	44,640	3360	480
39	Corfu, Otranto . . .	427,800	600	11,700	5880	420	60
40	Toulon, Corsica . . .	189,150	1950	39,000	18,135	1365	195
41	Malta, Alexandria . .	5,829,930	27,630	10,745	532,645	10,745	1535
42	Wexford	687,204	756	36,288	23,436	1764	63
43	England, Holland . .	2,439,840	1360	110,976	78,336	544	136
44	Sardinia, Sicily . . .	223,100	2300	42,400	36,000	1610	230
45	Persian Gulf	9,677,544	17,988	357,500	292,500	1499	1499

F.

SUBMARINE TELEGRAPH CABLES

Now in successful Working Order, the Insulated Wires for which were manufactured by the Gutta Percha Company, Patentees, Wharf Road, City Road, London.

No.	Date when Laid.	From	To	No. of Conductors.	Length of Cable in Statute Miles.	Length of Insulated Wire in Statute Miles.	Depth of Water in Fathoms.	By whom Covered and Laid.	Length of time the Cables have been working.
1	1851	Dover	Calais . . .	4	27	108	.	{ Wilkins & Wetherley, Newall & Co., Kuper & Co., and Mr. Crampton. }	14 years
2	1853	{ Denmark, across the Belt }	. . . . .	3	18	54		R. S. Newall & Co. . .	12 ,,
3	1853	Dover	Ostend . . .	6	80½	483		{ Newall & Co., and Kuper & Co. . . }	12 ,,
4	1853	Frith of Forth .	. . . . .	4	6	24	.	R. S. Newall & Co.	12 ,,
5	1853	Portpatrick .	Donaghadee .	6	25	150	.	,, ,, .	12 ,,
6	1853	Across River Tay.	. . . . .	4	2	8		,, ,,	12 ,,
7	1854	Portpatrick .	Whitehead . .	6	27	162	.	,, ,,	11 ,,
8	1854	Sweden . . .	Denmark . .	3	12	36	14	Glass, Elliot, & Co.	11 ,,
9	1854	Italy . . .	Corsica . . .	6	110	660	325	,, ,, .	11 ,,
10	1854	Corsica . . .	Sardinia . .	6	10	60	20	,, ,, .	11 ,,
11	1855	Egypt	. . . . .	4	10	40	.	,, ,, .	10 ,,
12	1855	Italy	Sicily . . .	3	5	15	27	,, ,, .	10 ,,
13	1856	Newfoundland .	Cape Breton .	1	85	85	360	,, ,, .	9 ,,
14	1856	{ PrinceEdward's Island . . }	{ NewBrunswick . . }	1	12	12	14	,, ,, .	9 ,,
15	1856	Straight of Canso.	{ Cape Breton, N.S. }	3	1½	4½	.	{ Nova Scotia Electric Telegraph Co. }	9 ,,
16	1857	Norway . across	Fiords . . .	1	49	49	300	Glass, Elliot, & Co. .	8 ,,
17	1857	{ Across mouths of Danube . }	. . . . .	1	3	3	.	,, ,, .	8 ,,
18	1857	Ceylon	{ Mainland of India }	1	30	30	.	,, ,, .	8 ,,
19	1858	Italy . . .	Sicily . . .	1	8	8	60	,, ,, .	7 ,,
20	1858	England . .	Holland . . .	4	140	560	30	,, ,, .	7 ,,
21	1858	Ditto . . .	Hanover . .	2	280	560	30	,, ,, .	7 ,,
22	1858	Norway . across	Fiords . . .	1	16	16	300	,, ,, .	7 ,,
23	1858	South Australia .	King's Island .	1	140	140	45	W. T. Henley . .	7 ,,
24	1858	Ceylon . . .	India	1	30	30	45	,, ,, .	7 ,,
25	1859	Alexandria . .	. . . . .	4	2	8	.	Glass, Elliot, & Co. .	6 ,,
26	1859	England . .	Denmark . .	3	368	1104	30	,, ,, .	6 ,,
27	1859	Sweden . . .	Gotland . . .	1	64	64	80	,, ,, .	6 ,,
28	1859	Folkestone .	Boulogne . .	6	24	144	32	,, ,, .	6 ,,
29	1859	{ Across rivers in India . }	. . . . .	1	10	10	.	,, ,, .	6 ,,
30	1859	Malta	Sicily . .	1	60	60	79	,, ,, .	6 ,,
31	1859	England . .	Isle of Man .	1	36	36	30	,, ,, .	6 ,,
32	1859	Suez	Jubal Island .	1	220	220	.	R. S. Newall & Co.	6 ,,
33	1859	Jersey . . .	Pirou, France .	1	21	21	15	Glass, Elliot, & Co. . .	5 ,,
34	1859	Tasmania . .	Bass Straits .	1	240	240	.	W. T. Henley .	5 ,,
35	1860	Denmark . . .	{ (GreatBelt) 14 miles 14 miles }	6) 3)	28	126	18	,, ,, .	5 ,,
36	1860	Dacca . . .	Pegu	1	116	116	.	,, ,, .	5 ,,
37	1860	Barcelona . .	Mahon . .	1	180	180	1400	,, ,, .	5 ,,
38	1860	Minorca . .	Majorca . . .	2	35	70	250	,, ,, .	5 ,,
39	1860	Iviza . . .	Majorca . . .	2	74	148	500	,, ,, .	5 ,,
40	1860	St. Antonio .	Iviza	2	76	152	450	,, ,, .	5 ,,
41	1861	Norway . across	Fiords . . .	1	16	16	300	Glass, Elliot, & Co. . .	4 ,,
42	1861	Toulon . . .	Corsica . . .	1	195	195	1550	,, ,, .	4 ,,
43	1861	Holyhead . .	Howth, Ireland	1	64	64	.	{ Electric & International Tel. Co. . }	4 ,,

APPENDIX.

SUBMARINE TELEGRAPH CABLES—*continued*.

No.	Date when Laid.	From	To	No. of Conductors.	Length of Cable in Statute Miles.	Length of Insulated Wire in Statute Miles.	Depth of Water in Fathoms.	By whom Covered and Laid.	Length of time the Cables have been working.
44	1861	Malta . . .	Alexandria . .	1	1535	1535	420	Glass, Elliot, & Co. .	3½ years
45	1861	Newhaven . .	Dieppe . . .	4	80	320	.	W. T. Henley, *laid* .	4 ,,
46	1862	Pembroke . .	Wexford . .	4	63	252	58	Glass, Elliot, & Co.	3¼ ,,
47	1862	Frith of Forth .	. . . . .	4	6	24	.	{ Electric & International Tel. Co. . }	3 ,,
48	1862	England . .	Holland . . .	4	130	520	30	Glass, Elliot, & Co.	2¾ ,,
49	1862	Across River Tay.	. . . . .	4	2	8	.	{ Electric & International Tel. Co. . }	3 ,,
50	1863	Sardinia . .	Sicily . . .	1	243	243	1200	Glass, Elliot, & Co.	2 ,,
51	1864	Persian Gulf .	. . . . .	1	1450	1450	120	{ W. T. Henley and Indian Government }	1 year
52	1864	Otranto . . .	Avlona . . .	1	60	60	569	W. T. Henley .	9 mths.
53	1865	La Calle . .	Biserte . . .	1	97¼	97¼	.	Siemens Brothers .	3 ,,
54	1865	Sweden . . .	Prussia . . .	3	55	166	.	W. T. Henley . .	1 month
55	1865	Biserte . . .	Marsala . . .	1	164¾	164¾	.	Siemens Brothers .	1 ,,

A great many Cables of short lengths, not included in this list, are now at work in various parts of the world ; and other Cables, the Wires insulated by the Gutta Percha Company, have been laid by Messrs. Felten & Guilleaume, of Cologne, during the last eight years, amounting to over 1000 miles, and which are now in working order.

G.

ATLANTIC TELEGRAPH COMPANY.

Report of the Directors to the Extraordinary General Meeting of Shareholders, held at the London Tavern, Bishopsgate Street, on Thursday, the 14th day of September, 1865.

12, St. Helen's Place, London,
13th September, 1865.

The sensation immediately consequent upon the recent accident to the Atlantic Telegraph Cable was one of profound disappointment, but this has to a great extent disappeared before the important and encouraging facts which were found to have been brought to light and practice during the expedition.

Not only has the future permanence of Deep-sea Cables been much enhanced by the greater convenience and safety with which they can be coiled and tested and payed-out since the Great Eastern has shown herself so well adapted to the work, but it has now also been proved absolutely that in the event of injury to the insulation, even after submersion, and while sunk in the deepest water, electricians are enabled with ease to calculate minutely the exact distance of the injured spot from ship or shore in a Cable 2,300 miles long.

It has further been proved that many miles of a Cable like that selected by the Atlantic Telegraph Company can, if so injured, be hauled in and repaired during the heaviest weather and from water 2000 fathoms in depth : and still more that even when a Cable is absolutely fractured, and the broken end lies at the bottom of an ocean 2000 fathoms deep, it is perfectly possible to find it and to raise it, and equally possible, according to the opinions of all those engaged in the recent expedition, to bring up the end of the Atlantic Cable, which is in that situation, and to splice it to the Cable on board the Great Eastern, so as to complete the communication to Newfoundland, so soon as apparatus of suitable strength and convenience can be manufactured.

In fact, so important have been the results of the last expedition in moderating every element of risk attendant on these undertakings, that the successful Submersion of submarine Cables will henceforward take its place as an event insurable for a moderate premium by the Underwriters.

The Directors, after careful investigation, therefore have determined not to relax in striving to bring to a successful issue the great work entrusted to their charge, but to press forward in the path of experience with increased vigilance and perseverance.

They have been encouraged in this view by the fair manner in which they have been met by the Contractors, with whom they have already entered into a contract for renewed operations.

Under this contract the Telegraph Construction and Maintenance Company undertake for the sum of 500,000*l.*, which has been agreed on as the cost price, at once to commence the manufacture of and during 1866 to lay down, a new Cable between Ireland and Newfoundland.

The Contractors, if the said Cable be successful, but not otherwise, are to have, in shares and cash, a profit at the rate of 20 per cent. upon such cost.

The Contractors also undertake, without any further charge whatever, to go to sea with sufficient Cable, including that now left on board the Great Eastern, and all proper appliances and apparatus such as experience has shown to be necessary, and to use their best endeavours—in the success of which they entirely believe—to recover and repair and complete in working order between Ireland and Newfoundland, the present broken Cable.

It will be seen that circumstances have thus enabled the Board to effect a very considerable economy in the Company's present operations.

It would no doubt have been a most gratifying circumstance if the recent accidents had not happened, and to the Directors this occurrence has been a grievous disappointment, but the circumstances surrounding the expedition and the increased confidence which, in spite of temporary discomfiture, has been given to the future of Deep-sea Cables, has enabled the Board to effect a new contract for the repair of the old Cable and for the submersion of a new one during 1866, on terms so satisfactory that if both these operations should succeed, the Company will actually be in possession of two efficient Cables for a less amount by 100,000*l.* than they would have been obliged to expend if the Cable of this year had been successful and the second Cable had been required to be purchased separately.

But the carrying out of this contract, so advantageous to the Atlantic Telegraph Company, involves the strenuous efforts of the Directors to raise an amount of money ranging from a minimum of 250,000*l.* to a maximum of 500,000*l.* in cash.

It is impossible that the Great Eastern ship could go to sea again this year to mend the existing Cable, and therefore such an operation, as a separate adventure, must be put out of the question, and even if undertaken separately would in itself involve an expenditure of some 120,000*l.*, whereas for a sum of 500,000*l.* the Contractors are willing to make and lay a new Cable next year in addition to the restoration of the old one ; they depending entirely upon success for profit.

The question which has had to be considered by the Directors in the interest of the Shareholders has been how best they might be enabled to raise this money.

The Eight per Cent. Preference Shares, though far below their real value, stand at 2*l.* 5*s.* per share, and if the Company were to adopt the alternative of winding-up its affairs, their intrinsic worth would not be 10*s.* per share.

The expenditure of the new money will certainly create fresh property, and probably resuscitate the old.

By its means the existing Eight per Cent. Preference Stock will doubtless be placed at par in the market before the sailing of the ship next year.

The Directors are, however, compelled to offer an inducement to those who are willing to come in and assist to place in that position the Company's, at present, sinking property.

Acting under advice, and believing in the very large profits that undoubtedly await this Company when successful, they desire to offer a first dividend of 12 per cent., with participation in profits, after 8 per cent. has been paid upon the existing preference shares and 4 per cent. upon the old capital, to those who consent to supply the requisite funds.

The Shareholders will have the opportunity of subscribing for this new Preferential Stock, which is issued solely to protect their property. Those proprietors who subscribe to it are manifestly not injured in any way, as they absorb the whole profits of the Company. Those who do not subscribe pay in effect a small premium to the subscriber who comes forward to help them. It is considered by the Board that this is infinitely preferable to winding-up the Company, whereby the Shareholders would have the mortification of seeing the whole of their property sacrificed, and of seeing an undertaking pass out of their hands, when on the very eve of success, upon which so much attention has been bestowed, and so much experience gained by the expenditure of their own funds.

Such a sacrifice is totally unnecessary, for it can be ascertained by any one who will take the trouble to make a small calculation, that if each of the two proposed Cables can be worked at the very low rate of only five words per minute upon each Cable for sixteen hours a day at five shillings per word, which is believed to be a much lower rate than the pressure of business would admit of in the first instance, the traffic, after paying the dividend charges of 12, 8, and 4 per cent. respectively, amounting together to 144,000*l.* upon the capital comprised in those different stocks, and after adding thereto the very large sum of 50,000*l.* a-year for working expenses, would leave an enormous balance for paying further dividends or bonuses on the Company's total capital, both ordinary and preferential.

BRADBURY, EVANS, AND CO., PRINTERS, WHITEFRIARS.

Printed in the United States
By Bookmasters